FIRING
CERAMICS

ADVANCED SERIES IN CERAMICS

Editors-in-Chief: M McLaren and D E Niesz

Advanced Series in Ceramics – Vol. 2

FIRING CERAMICS

G. Bickley Remmey, Jr.

World Scientific
Singapore • New Jersey • London • Hong Kong

Published by

World Scientific Publishing Co. Pte. Ltd.
P O Box 128, Farrer Road, Singapore 9128
USA office: Suite 1B, 1060 Main Street, River Edge, NJ 07661
UK office: 73 Lynton Mead, Totteridge, London N20 8DH

Library of Congress Cataloging-in-Publication Data

Remmey, G. Bickley.
   Firing ceramics / G. Bickley Remmey.
     p.   cm. -- (Advanced series in ceramics ; vol. 2)
   Includes index.
   ISBN 9810216785. -- ISBN 9810216793 (pbk.)
   1. Kilns.   2. Ceramics.   I. Title.   II. Series.
TP841.R45   1994
666'.443--dc20   94-21247
                                           CIP

Printed in Singapore.

PREFACE

Most existing literature covering ceramic manufacturing processes cover the entire spectrum from raw material to finishing. These texts will often have only one or two chapters devoted to the firing process. Therefore, this book has been written totally on the firing process to serve not only as an in depth textbook for our ceramic schools, but as a firing handbook for anyone who manufactures ceramics.

This book is divided into three parts with Part I describing what happens inside a kiln to the ceramic itself, plus what kiln furniture may be required and how to develop a firing cycle. Part II deals with all the different kinds of kilns that exist and how to select the right kind for your job. Lastly, Part III is devoted to the latest in firing practice, industry by industry.

ACKNOWLEDGEMENTS

I would like to express my appreciation to those who have helped me with this project. First of all to my wife Jeri for her encouragement and support throughout my entire career. I would like to thank Swindell Dressler Company for their assistance in preparing many of the illustrations as well as the use of numerous kiln photographs. I would like to thank Edward F. Howe, Russell K. Wood, Gerald A. Wagner, Robert Lys and Ronald Latham for their comments on firing practice in their respective industries. Lastly, I would like to thank Dr Malcolm G. McLaren and Dr Dale E. Niesz of Rutgers, plus Andrew M. Halapin, James D. Bushman and James G. Hopkins for reviewing the text and making valuable contributions prior to the submittal to the publishers.

DEDICATION

To my father who taught me the kiln business and who made so many significant advances in the field.

CONTENTS

PART I
THE FIRING PROCESS

1. FIRING REACTIONS

The firing process can be defined as the process where ceramic powders and/or clay, which have been compacted, are heated to a temperature where useful properties will be developed.

The firing process encompases chemical and physical changes in the ceramic body accompanied by a loss of porosity and a subsequent increase of density. The compacted powder body becomes bonded together in a rigid matrix by vitrification which involves glass formation or by sintering where little or no liquid is present. The major ceramic product groups and their primary bonding systems are shown in the following chart.

Type Ceramic Product	Primary Bonding System
Structural Clay	
Whiteware	
Fireclay Refractories	Vitrification
Abrasives	
High Alumina	
Basic Refractory	
Ferrites	Sintering – Solid State
Pure Oxides	Sintering – Liquid Phase
Pure Carbides	

1.1 Vitrification

Vitrification by definition is the progressive reduction and elimination of porosity of a ceramic composition with the formation of a glass phase as a result of heat treatment. Glass formation typically starts around 1100°C and accelerates with further increase in temperature. The amount of glassy phase that is allowed to develop varies from one ceramic product to the next, based on the properties desired in the product.

A fireclay refractory brick may have as little as 8% glassy phase and a translucent fine china may have more than 60% glassy phase. However, in all cases, the rate of glass formation increases with higher temperature or more time at temperature.

Once the desired amount of porosity is obtained, the cooling is started. During cooling, the glassy phase freezes and becomes rigid to form a strong bond with the crystalline phase of the body.

Illustration 1 shows a micrograph of an unfired porcelain body which is basically compacted powder of varying grain sizes with a large number of small open pores.

Illustration #1 – Micrograph – Unfired Porcelain. Courtesy of Rutgers University.

Illustration 2 shows a micrograph of fired porcelain with the glassy phase developed. Note pores are now closed instead of open, fewer in number, and much larger in size.

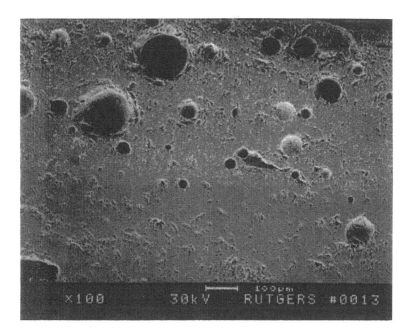

Illustration #2 – Micrograph – Fired Porcelain. Courtesy of Rutgers University.

During the vitrification process, the following physical changes take place in the ceramic body.

1. Shrinkage due to loss of open pores.
2. Development of closed pores.
3. Development of glassy phase.

These physical changes are shown in Illustration 3 which charts what happens to a typical porcelain on firing. Note the open pores are reduced from 33% to almost nothing. The closed pores, which start developing around 1100°C, end up being about 7% of the fired volume. The glassy phase, in this case, becomes 57% of the fired volume.

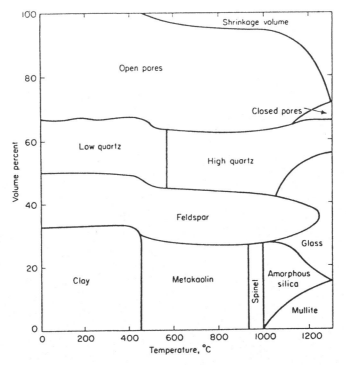

Illustration #3 – Chart Showing What Happens to a Porcelain on Firing. Courtesy of Fine Ceramics – Norton.

1.2 Sintering

Sintering by definition is a process of permanent chemical and physical change accompanied by reduced porosity by the mechanism of grain growth and grain bonding. Sintering temperatures vary depending on the material being sintered. Some ferrites can be sintered as low as 1250°C whereas the sintering temperature for 99% alumina is 1800°C.

During the sintering process, the large grains grow at the expense of the small grains so as the average size of the grains increases, the number of grains decreases. The mechanisms for sintering are:

Solid State Sintering

1. Material transfer by vapor transport between particles.
2. Material transfer by solid state diffusion.

Liquid Phase Sintering

1. Material transfer by solubility and precipitation in the liquid phase.

Illustration 4 shows a schematic of vapor-phase material transport. This mechanism bonds the particles together to produce a larger particle but does not cause shrinkage.

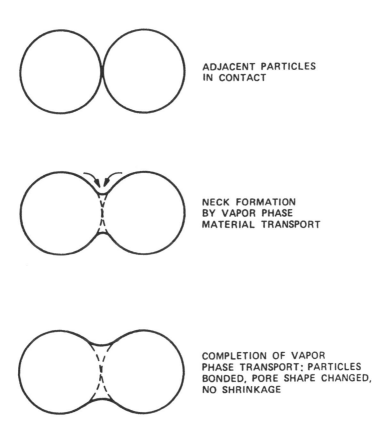

ADJACENT PARTICLES
IN CONTACT

NECK FORMATION
BY VAPOR PHASE
MATERIAL TRANSPORT

COMPLETION OF VAPOR
PHASE TRANSPORT; PARTICLES
BONDED, PORE SHAPE CHANGED,
NO SHRINKAGE

Illustration #4 – Schematic Vapor-Phase Material Transfer.

Illustration 5 shows a schematic of solid-state material transport by diffusion. Diffusion is the movement of atoms sometimes called atom jump along a grain boundary or through the grain itself which is called lattice diffusion. Lattice diffusion does cause shrinkage.

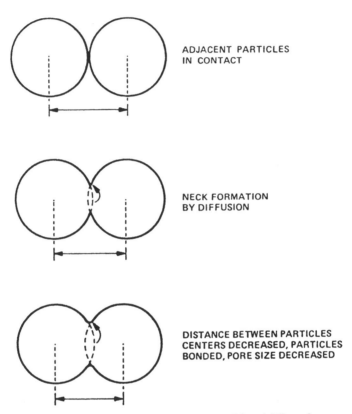

ADJACENT PARTICLES
IN CONTACT

NECK FORMATION
BY DIFFUSION

DISTANCE BETWEEN PARTICLES
CENTERS DECREASED, PARTICLES
BONDED, PORE SIZE DECREASED

Illustration #5 – Schematic – Solid State Material Transfer.

Liquid Phase Sintering

Liquid phase sintering or sintering in the presence of a reactive liquid also leads to densification. This occurs in systems where the solid phase shows a certain limited solubilitiy in the liquid at sintering temperatures. The mechanism of this sintering process is the solution and reprecipitation of the solids to give increased grain size and density. Some ceramics where liquid phase sintering occurs are: alumina substrates, spark plug insulators, titanates, and hard ferrites.

During the sintering process, the sintering rate increases with increased temperature level. The rate of sintering also increases with reduced particle size. Lastly, as densification approaches maximum levels, the rate of sintering decreases with time.

Illustration 6 shows a micrograph of an unfired, high alumina body which at this stage is compacted alumina powder of varying grain sizes with a large number of open pores.

Illustration #6 – Micrograph – Unfired Alumina. Courtesy of Rutgers University.

Illustration 7 shows a micrograph of a sintered high alumina body. Note that the grains have grown in size and bonded together. Most of the pores are now closed.

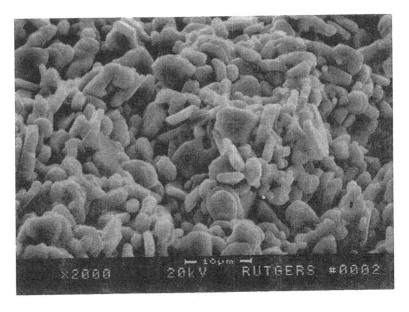

Illustration #7 – Micrograph – Fired Alumina. Courtesy of Rutgers University.

1.3   Other Reactions During Firing

Besides vitrification and sintering, a number of other reactions also take place during firing. These reactions can be grouped into the following categories:

A. Loss of physical water
B. Oxidation
C. Decomposition
D. Quartz transformations

A. *Loss of Physical Water* (100–200°C)

Most green ceramics contain 1 to 3% physical water when they are placed in the kiln. When the firing process starts, this physical water comes off as water vapor between 100 and 200°C which in essence completes the drying process.

It is important to note here that a kiln makes an expensive dryer so that products that are formed with a high percentage of water (extruded structural clay products can have 18% water) should be dried down to less than 1% before placing in a kiln.

The reason for this is that the drying process can take as much time as the firing process so that if you dried the product in a kiln, you would have twice as many kilns. Since a dryer costs less to build than a kiln, it makes economic sense to have a separate dryer.

B. *Oxidation* (200–800°C)

Organic materials typically burn out between 300–400°C. Organic material is often added to green ceramic bodies as a temporary binder. The purpose of the temporary binder is to hold the green ceramic together so it can be handled and placed in a kiln. Another source of organic material in a green body are materials such as saw dust which is added to the body so it will burn out to create porosity. Still another source of organic is the naturally occurring organics in ball clay.

The carbon in these organics oxidizes to carbon dioxide ($C + O_2 \longrightarrow CO_2$ which is an exothermic reaction. In cases where a large amount of organic is present, it is sometimes necessary to go through the oxidation period with a relatively low oxygen level in the kiln. This slows down the oxidation process and prevents the ceramic from cracking due to excessive heat generated inside the piece by the CO_2 formation. In this case, a two-stage reaction occurs with formation of carbon monoxide ($2C + O_2 \longrightarrow 2CO$) inside the piece which releases less heat than ($C + O_2 \longrightarrow CO_2$). Once the CO gas volatilizes out of the ceramic it then completes the oxidation to CO_2 in the kiln atmosphere and or flue which does not harm the ceramic.

It is important to oxidize all the carbon in a ceramic body and volatilize it out of the ceramic through the open pores before densification starts to close the pores. When this situation occurs, black coring will occur with pure carbon trapped inside the piece. Another possibility is bloating caused by CO or CO_2 trapped inside the piece by closed pores. To prevent black coring and bloating it may be necessary to slow down the firing rate through the oxidation range.

The higher the percentage of organic, the slower the firing rate. Also the larger the size of the ceramic piece, the slower the firing rate.

Besides organics, sulfides also will oxidize during the firing process. Sulfides burn out between 380–800°C to form SO_2 gas.

C. *Decomposition*

The dehydroxylation of clay sometimes called "the loss of chemical water" occurs between 480–700°C. The decomposition of Kaolin to Meta-Kaolin occurs by the reaction.

$$Al_2 \ Si_2 \ O_4 \ (OH)_4 \longrightarrow Al_2 \ Si_2 \ O_6 + 2 \ H_2O$$

Some other materials and their decomposition temperatures are as follows:

- Hydrates   Decompose between 100–1000°C giving off H_2O
- Carbonates   Decompose between 400–1000°C giving off CO_2
- Sulfates   Decompose at 1000–1200°C giving off SO_2. Sulfates are stable as $BaSO_4$ and $CaSO_4$ in products fired below 1200°C
- Meta-Kaolin   Decomposes between 1000–1200°C to mullite and silica
- Kyanite   Decomposes between 1300–1450°C to mullite and silica with an increase in volume

D. *Quartz Transformations*

Silica goes through several polymorphous modifications on heating which are as follows:

below 573°C   alpha quartz or low quartz
573–867°C   beta quartz or high quartz
867–1470°C   tridymite
1470–1710°C   cristobalite
above 1710°C   liquid

Any cristabalite or tridymite formed will remain on cooling but the majority of silica grains in most ceramics never transform beyond Beta quartz. When cooled below 600°C, beta quartz transform abruptly back to alpha quartz with a sudden volume change. Slow cooling between 600–500°C is often required to prevent cracking. The larger the piece and the greater the silica content, the slower the cooling must be during this transformation. The sudden volume change of alpha to beta quartz on heating at 573°C does not cause as much strain in the green ceramic body as cooling; however, heating cracks due to the quartz inversion can occur in high silica bodies or in large ceramic pieces.

Because of the alpha to beta quartz transformation, alumina is often substituted for silica in fast fire bodies.

2. EXHAUST EMISSIONS

Prior to the Clean Air Act of 1970, exhaust emissions were simply dumped into the air in the U.S.A. The common practice was to adjust the firing cycle time so binder burnout would occur at night and no one would notice the smoke coming from the kiln. Today, however, with strict anti-pollution laws, emissions are no longer permitted beyond specified safe limits. The three most common forms of pollution that can come from a kiln are combustibles in the form of smoke, sulphur in the form of SO_2 and fluorine in the form of HF.

2.1 Combustibles

Far and away the most common type of pollutants coming out of kilns are combustibles. The combustibles are carbon, carbon monoxide and hydrogen; all of which are a result of the organics volatilizing out of the ceramic bodies. In order to meet air pollution standards, it is necessary to complete the oxidation process and convert the combustibles to carbon dioxide and water which are not pollutants:

$$C + O_2 \longrightarrow CO_2$$
$$2CO + O_2 \longrightarrow 2CO_2$$
$$4H + O_2 \longrightarrow 2H_2O$$

Unfortunately, it is not possible to perform the above reactions in the kiln because the combustibles come off in the form of smoke in

14

the 300–400°C range; however, the oxidation to CO_2 and H_2O does not occur until 650–700°C is reached. Therefore, the most common way to eliminate the combustibles is with an after burner built into the flue or stack. The afterburner heats the 300–400°C exhaust gas in the presence of oxygen up to 650–700°C so that the oxidation reactions can occur.

Illustration 8 shows an afterburner system in a typical downdraft type batch kiln. As can be seen in the illustration, smoke coming from the kiln at 300–400°C is heated to 650–700°C by the afterburner located in the flue. The combustibles oxidize in the refractory lined flue. A rule of thumb when designing afterburner systems is to provide enough cross section in the flue to provide 0.5 seconds of dwell time in the afterburner zone. Afterburners of this type are typically 99.9% efficient in converting all combustibles.

In updraft batch kilns and tunnel kilns, the afterburner is built into a refractory lined portion of the exhaust stack.

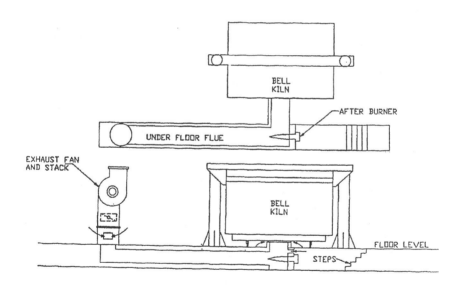

Illustration #8 – Afterburner System.

2.2   Sulphur

Sulfides and some clays contain sulphur which volatilize to sulphur dioxide (SO_2) in the kiln. Since SO_2 is a major contributor to acid rain, it must be eliminated from the kiln exhaust. The most common way to handle SO_2 is with a packed tower type wet scrubber as shown in Illustration 9.

The packed bed is usually stainless steel which is basically a maze for the gases to pass through which provides lots of surface area for the gases to come in intimate contact with the scrubbing liquid. The scrubbing liquid is water with sodium hydroxide (NaOH) dissolved in it. The NaOH reacts with the SO_2 to form $NaSO_4$ which is carried away by the water.

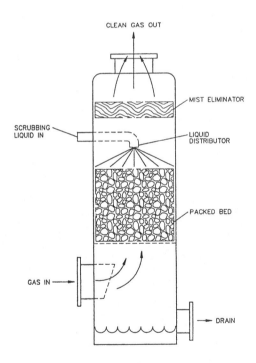

Illustration #9 – Sulphur Scrubber.

2.3 Fluorine

Certain ceramic raw materials such as clay and talc can contain Fluorine. During the firing process the Fluorine evolves in the form of HF (HydroFluoric acid gas) between 800–1100°C. The amount of HF gas generated often exceeds what is legally permitted so that a fluorine absorber is required. Fluorine absorbers work on the principle of pulling the exhaust gases through beds of limestone granules which absorb the Fluorine to create Fluorspar.

Illustration 10 shows a large Fluorine absorber where the Fluorine is removed after which, the Fluorine free exhaust is discharged into the brick lined stack and into the atmosphere.

Illustration #10 – Fluorine Absorber.

2.4 NO_x (Nitrogen and Oxygen Compounds)

NO_x, which is the symbol for nitrogen and oxygen compounds as a group, has been identified as one of the pollutants that can cause acid rain. Limitations on NO_x emissions have already been specified in some countries including the USA. The amount of NO_x in the exhaust from a kiln increases with flue gas temperature and also increases with oxygen level in the kiln. Since flue gas temperature is a function of the firing process, it cannot normally be changed. However, the amount of oxygen in the kiln can be controlled.

Therefore, the primary method of controlling NO_x emission today is to maintain low oxygen levels (2%) throughout the higher temperature portions of the firing cycle. The most effective firing system for controlling a kiln at low oxygen levels is the pulse firing system which will be discussed in Part II of this book.

3. KILN ATMOSPHERE

3.1 Oxidizing Atmosphere

The great majority of all ceramic products are made up of oxides and are fired in an oxidizing atmosphere. Since the combustion process does not produce any free oxygen, $CH_4 + 2O_2 \longrightarrow CO_2 + H_2O$, an oxidizing atmosphere is created by adding a certain amount of excesss air. The following chart, which is based on burning natural gas, shows the oxygen content in the kiln atmosphere with varying amounts of excess air.

	% Excess Air	% Free Oxygen
Stoichiometric (no excess air)	0%	0%
	50%	7.3%
	100%	10.8%
	1000%	19.1%

The ability to control oxygen level in the firing process is very important and this is one of the reasons for the great popularity of excess air burner systems today. Obviously, control of the oxygen level effects the oxidation process and can be used to control the rate of binder burnout. Also, it is desirable to maintain some oxygen level (2 to 3%) throughout the soak period (holding period at maturing temperature) to prevent reduction of oxides in the ceramic body.

3.2   Reducing Atmosphere (CO + CO$_2$)

Some products such as electrical porcelain made by the European process and clay-graphite refractories require a reducing atmosphere. It is interesting to note that electrical porcelain fired in the USA is fired in an oxidizing atmosphere but most of the electrical porcelain fired in the rest of the world is fired under reducing conditions. The typical firing cycle for the reduction firing of insulators is oxidizing up to approximately 950°C; from 950°C to soak temperature, (1300–1350°C) it is reducing and then the atmosphere changes back to oxidizing for cooling. The reduction period starts out with a very strong reducing atmosphere of 4–5% CO and then is changed at about 1300°C to 1–1.5% CO.

Clay-Graphite refractories must be kept reducing above 650–700°C to prevent the carbon from burning out of the product to CO and CO$_2$. Once the vitrified bond develops and the pores are closed, the carbon loss stops and the product can be cooled in an air atmosphere.

3.3   Nitrogen Atmosphere

Some products such as soft ferrites and nitride bonded silicon carbide are fired in a nitrogen atmosphere, but for totally different reasons.

Soft ferrites (ceramic electromagnets) are fired in gas tight electric kilns where nitrogen is used as a protective atmosphere to protect the product from oxygen.

Especially important is the cooling part of the soft ferrite firing cycle which must be done in an oxygen free environment to obtain correct magnetic properties.

The situation with nitride bonded silicon carbide is totally different. In this product, the nitrogen actually enters into the ceramic body as a bond with the silicon carbide matrix. The fired product can be 8% by weight of nitrogen. This product is fired inside of a silicon carbide muffle which is pressurized on the inside with nitrogen gas. The atmosphere outside the muffle is oxidizing, however, the pressure inside the muffle is kept higher than the ambient kiln pressure so that there is, in essence, a slow leak out but no leak into the muffle.

4. KILN FURNITURE

4.1 Purpose

Kiln furniture is a general term describing special refractory shapes that are designed to support green ceramics during the firing process. All products that cannot be stacked on top of each other for firing such as brick, or cannot be fired in a roller hearth kiln directly on the rolls such as ceramic tile, will require kiln furniture.

Kiln furniture can weigh more than the product itself so that the type of kiln furniture being used can strongly effect the firing cost. Since it costs just as much to fire a kilogram of kiln furniture as it does to fire a kilogram of product, it is desirable to minimize the weight of kiln furniture wherever possible.

The most common materials used for kiln furniture today are as follows:

	Max. Use Temperature
Cordierite	1275°C
Mullite	1750°C
Silicon Carbide	1500°C
90–99% Alumina	1750°C
Zirconia	1650°C

Cordierite is the least expensive type kiln furniture and is the most widely used below 1300°C. Cordierite is also relatively light in weight and has almost no thermal expansion and therefore, excellent thermal shock resistance. This is extremely important to firing cost since the life of kiln furniture is measured in number of heating and cooling cycles. Well designed cordierite kiln furniture can last many hundreds of cycles.

Mullite is several times more expensive than Cordierite and can be used to temperatures as high as 1750°C. Mullite is heavier than Cordierite but is very strong and, therefore, can be made with thinner cross sections. Mullite has good thermal shock resistance due to its high strength.

Silicon Carbide (SiC) has very high tensile strength and the direct bonded version of this material can be made into load bearing rectangular hollow structural tubing capable of spanning several meters. This structural tubing, due to the hollow cross-section, is also lightweight. Direct bonded silicon carbide can be used up to 1600°C. This tubing is very useful in supporting shelving, however, it is quite expensive.

Standard silicon carbide, which is less expensive than the direct bonded type, is also commonly made into batts or setter tiles to be used in shelving. The advantage of this material as shelving is strength, high thermal conductivity, good thermal shock resistance, and use temperatures to 1500°C.

The two types of high alumina kiln furniture are 90% (use temperature 1650°C) and 99% (use temperature 1800°C). Alumina is a little more expensive than mullite, however, it can be used for temperatures as high as 1800°C. Like mullite, it has good thermal shock resistance due to its high strength and can be used for fast cycles (90 min.).

Zirconia is also made into kiln furniture for use up to 1650°C. It's major use is where 99% alumina cannot be used for chemical reasons.

4.2   Kiln Furniture Systems

Some of the most common types of kiln furniture systems are as follows:

A. *Structural Clay Products (Illustration #11)*

This kiln furniture system uses vertical, hollow round cordierite tubes to support cordierite plates with holes which form the shelves. This kiln furniture system is lightweight and very thermal shock resistant which translates into low fuel cost and long life.

Illustration #11 – Kiln Furniture for Structural Clay Products. Courtesy of Ferro Corporation.

B. *Dinnerware* (*Illustration #12*)

This kiln furniture system uses cordierite cranks which is especially designed kiln furniture to support glazed dinnerware without contacting the glaze.

Illustration #12 – Kiln Furniture For Dinnerware. Courtesy of Ferro Corporation.

C. *Tile* (*Illustration #13*)

The tile Industry uses a variety of cordierite designs as shown in Illustration #13. Kiln furniture for glazed tile support each tile separately.

Illustration #13 – Kiln Furniture for the Tile Industry. Courtesy of Ferro Corporation.

D. *Electronic Ceramics* (*Illustration #14*)

Powders and small parts are often fired in saggers (refractory containers). Saggers can be round or rectangular; however, the most popular are rectangular since they stack more efficiently in a kiln. The maximum size of the sagger is usually determined by the weight of the product since the loaded sagger must be a reasonable weight for a man to lift. The material saggers are made of is dependent on firing temperature. Illustration #14 shows a variety of sagger designs made of Mullite, Alumina and Zirconia.

Illustration #14 –Kiln Furniture for the Electronics Industry. Courtesy of Ferro Corporation.

Illustration #15 – Kiln Furniture For the Sanitaryware Industry. Courtesy of Ferro Corporation.

E. *Sanitaryware (Illustration #15)*

Kiln Furniture used in the Sanitary Ware Industry consist of Cordierite posts and support pedestals plus recrystalized SiC hollow structural tubing as shown in Illustration #15.

F. *Electrical Porcelain (Illustration #16)*

Illustration #16 shows cordierite adjustable shelving which can be adjusted to fire various size insulators.

Illustration #16 – Kiln Furniture for the Electrical Porcelain Industry. Courtesy of Ferro Corporation.

5. FIRING CYCLES

5.1   The Effect of Body Composition on the Firing Cycle

The body composition determines the ultimate firing temperature and also can effect the heating and cooling rates. Illustration #17 shows typical firing temperatures for various ceramic products. This chart is helpful in determining approximate soak temperature, however, other factors must be taken into consideration to determine soak time such as the setting arrangement and the size of the product being fired.

Pyrometric cones are very helpful in determining the length of soak time required at temperature. Pyrometric cones measure the effect of time and temperature together rather than just temperature alone like a thermocouple. Pyrometric cones are made of standardized unfired ceramic compositions which undergo the same changes during firing as your ceramic product body does. For example, the progressive formation of glass will eventually cause the cone to slump under its own weight so when the tip of the cone touches the cone plaque, the cone has reached its end point.

It is common practice to use three different cones for control, however, they can be used singly or in groups. A group would consist of a "firing cone" which is the desired cone for the ware being produced, the "guide cone", one cone cooler and the "guard cone" one cone hotter. They are set deep enough in the kiln so that all three cones are clearly visible through a peephole When the "guide cone" starts to bend you know the ware is approaching maturity. Then bending of the "firing cone" indicates firing is at the correct point. (See Illustration #18). If the "guard cone" is bent, you may have exceeded the best time-temperature relationship. When viewing the cones through the peephole, it is wise to wear a pair of dark sunglasses to avoid possible eye injury from the intense heat and light.

TYPICAL FIRING TEMPERATURES OF CERAMICS

	°C
ABRASIVES	1250-1300
ELECTRONIC & TECHNICAL CERAMICS	
cordierite	1300-1450
steatites	1260-1360
aluminas	1500-1760
titanates	1290-1370
ferrites	1200-1450
beryllia	1700-1850
zirconia	1700-1800
HEAVY CLAY PRODUCTS	
common brick	1000-1100
face brick	1050-1150
sewer pipe	1100-1150
roof tiles	1100-1150
clay tile	1000-1100
POTTERY	
stoneware	1270-1330
earthenware	1250-1300
artware	1000-1200
flower pots	900-1000
REFRACTORIES	
fireclay	1260-1400
high alumina	1450-1540
silica	1450-1510
chrome brick	1450-1650
magnesite brick	1450-1650
direct bonded basic	1700-1800
silicon carbide	1400-1510
WHITEWARE (once fired)	
electrical porcelain	1150-1260
sanitaryware	1200-1300
hotel china	1200-1260
fine china	1140-1200
floor tile	1100-1230
wall tile	1000-1230

Illustration #17 – Typical Firing Temperatures of Ceramics.

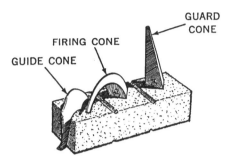

GUARD CONE

FIRING CONE

GUIDE CONE

Illustration #18 – Three Cone System. Courtesy of the Edward Orton, Jr. Foundation.

Temperature Equivalents
Table for Orton
Standard Pyrometric Cones

Cone Number	Temperature Equivalent (Note 1)	Approximate Color of Kiln Interior (Note 2)	Type of Ware and Glazes (Note 3)
15	2608°F		
14	2491		
13	2455		
12	2419		
11	2399	White	Porcelain
10	2381		
9	2336		China Bodies
8	2305		Stoneware
7	2264		Salt Glazes
6	2232		
5	2185		
4	2167		China Glaze
3	2134		
2	2124		Semi-vitreous ware
1	2109		
01	2079		Earthenware
02	2048	Yellow	
03	2014		
04	1940		Low Fire Earthenware
05	1915		Lead Glazes and
06	1830		Low Fire Fritted Glazes
07	1803		
08	1751	Orange	
09	1693		
010	1641		
011	1641		
012	1623		Lustre Glazes
013	1566	Cherry Red	
014	1540		
015	1479		Chrome Red Glazes
016	1458		
017	1377		Overglaze colors,
018	1323		Enamels and
019	1261		Gold
020	1175	Dull Red	
021	1137		
022	1090		

Note 1 The temperature equivalents in this table apply only to Large
(2 1/2") Orton Pyrometric Cones when heated at the rate of 270°F per
hour in an air atmosphere.

Illustration #19 – Orton Standard Pyrometric Cones. Courtesy of Edward Orton, Jr. Ceramic Foundation.

One technique for determining the length of soak time in a batch fired kiln is to place cones on the perimeter of the load and also in the center or bottom of the load which should be the slowest to heat. The kiln is heated until the cones on the perimeter of the load start to come down at which point the temperature is held and the soak period begins. When the cones in the center of the load start coming down, the soak period is finished and cooling can begin. Illustration #19 shows the temperature equivalents table for Orton standard pyrometric cones versus firing temperatures for various ceramics and glazes.

Body composition can effect the heating and cooling rates depending on what changes occur in the ceramic body with temperature. One of the most widely used means of observing these changes is with Differential Thermal Analysis (DTA). This test measures the exothermic and endothermic reactions that occur on heating a sample ceramic body at a constant rate (10°C per minute) as compared to an inert material (usually alumina). The types of reactions measured by a DTA test are as follows:

Endothermic:	decompositions
(absorb heat)	crystal transitions
Exothermic:	oxidations
(gives off heat)	new phase formations

Illustration #20 shows a DTA analysis for Kaolinite clay. The endothermic reaction is the decomposition to meta-Kaolin (loss of chemical water) between 500°C and 700°C. Between 950°C and 1000°C an exothermic reaction occurs which is a new phase formation (meta Kaolin to mullite).

In addition to DTA analysis, Thermal Gravimetric Analysis (TGA) and X-ray diffraction are also used to determine the nature of changes in a ceramic body during heating. TGA measures weight changes of a sample while being heated at a constant rate. X-ray diffraction is used to identify crystal forms in the body.

Illustration #21 shows what happens at various temperatures during the firing cycle as a result of body composition. These various chemical and physical changes can effect the shape of the heating and cooling curves as well as the heating and cooling rates.

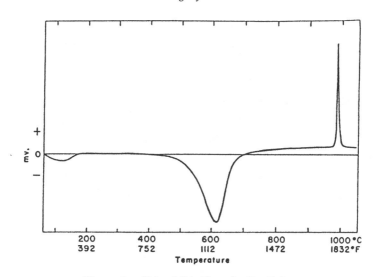

Illustration #20 – DTA Chart for Kaolinite.

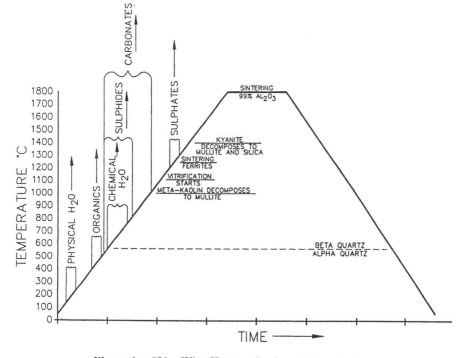

Illustration #21 – What Happens During a Firing Cycle.

Some examples of how the shape of the firing curve can be modified to accommodate chemical and physical changes occurring during firing are as follows:

1. *Physical Water (See Illustration #22)*

Physical water can be a problem with large shapes. A typical example of a firing cycle modification designed to remove physical water without steam generation in large cross section ceramic shapes is shown in Illustration #22. Here the temperature is held at 90°C for hours to allow time for the physical water to move by capillary action from the inside of the ceramic to the surface where it vaporizes to water vapor.

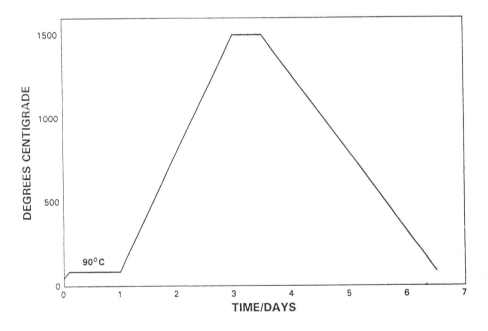

Illustration #22 – 24 Hours Hold at 90°C to Remove Physical Water from Large Shapes.

2. *Organic Burnout* (*See Illustration #23*)

Organic materials can be added to a green ceramic body as a burnout material to control porosity. The amount of organic can be quite substantial and since the burnout is an exothermic reaction, considerable heat is released. To control the burnout it is often necessary to slow down the heating rate during burnout as well as control the oxygen level to slow down the reaction. Illustration #23 shows slow heating from 200 to 500°C to control the organic burnout.

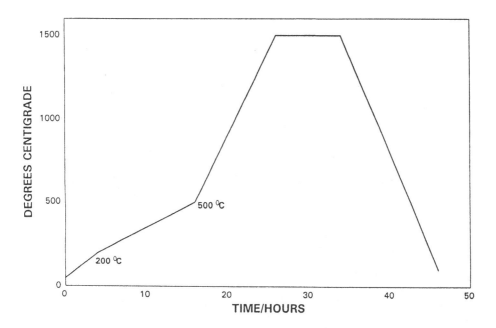

Illustration #23 – Slow Heating from 200–500°C to Control Organic Burnout.

3. *Quartz Transformations* (*See Illustration #24*)

The alpha to beta quartz transformation is accompanied by a large volume change which can crack a ceramic with high silica content especially during cooling.

Therefore, it is necessary when cooling to go through the inversion at 573°C with the entire piece at almost the same temperature. The larger the piece and/or the higher percent of silica in the body, the more severe is this problem. Illustration #24 shows a typical firing cycle for silica coke oven blocks. Note that the heating and cooling rates below 600°C are very slow whereas above 600°C the heating and cooling can occur at much faster rates without damage to the product.

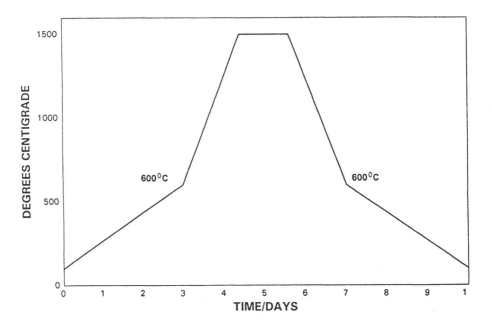

Illustration #24 – Silica Coke Oven Block Cycle with Slow Heating and Cooling Below 600°C for Quartz Inversions.

5.2   The Effect of Product Size on the Firing Cycle Time

The size of the product being fired effects the time coordinant of the firing curve. The larger the piece being fired the longer the firing cycle has to be. The reason for this is that the kiln can only deliver heat to the surface of the piece after which the heat is conducted from the surface to the interior. Since most ceramics are heat insulators, heat transfer by conduction is relatively slow.

Also, the rate of heat transfer by conduction is dependent on the thermal gradient between the surface and the interior. However, since ceramics are weak in tensile strength, if too large a thermal gradient is set up, differential expansion will occur between the surface of the piece and the interior causing the piece to crack.

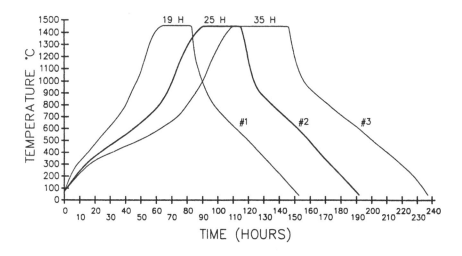

Illustration #25 – Firing Cycle for Various Size Silica Blocks.

Illustration #25 shows the effect of product size on firing cycle time. Firing curve No. 1 is for silica blocks less than 15 Kg, firing cycle No. 2 is for silica blocks up to 30 kg and cycle No. 3 is for blocks more than 30 kg. As can be seen, all three parts of the firing cycles (Heating – Soak – Cooling) are increased in time as follows:

	Heating Time (Hours)	Soak Time (Hours)	Cooling Time (Hours)	Total Time (Hours)
Cycle 1	65	19	66	150
Cycle 2	92	25	73	190
Cycle 3	110	35	98	240

When firing various size pieces in one kiln load, the firing cycle must be tailored to the largest piece. For this reason it often pays to segregate pieces by size and have more than one firing cycle. Another case in point is where a variety of sizes are being fired in a tunnel kiln on a firing cycle determined by the largest piece. Often these large pieces represent a small percentage of the total such as 15%. By pulling that 15% out of the tunnel kiln and firing them in a separate batch kiln, the tunnel kiln can often be speeded up to fire 30 to 50% more product per day.

5.3   The Effect of Setting Arrangement on Firing Cycle Time

Another factor that effects firing cycle time is the setting arrangement in the kiln. For example, if large grinding wheels 1 meter diameter × 50 mm thick are fired one high on kiln furniture shelving, the firing cycle might be 84 hours. If these wheels were stacked 3 high, the cycle might increase to 130 hours; however, the kiln would fire more wheels per week with less fuel per wheel and use less kiln furniture.

One of the important developments of recent years is the concept of one high firing in continuous kilns. In the case of wall tile, kiln furniture (cranks) used in conventional tunnel kilns firing 10 hour cycles are completely eliminated when going to one high firing in a roller hearth kiln with a one hour cycle. Other ceramics such as dinnerware and sanitaryware can be fired one high in car type tunnel kilns on significantly faster cycles than if the product were stacked using multiple layers of kiln furniture.

PART II
KILNS

6. KILN HISTORY

The earliest kiln was undoubtedly nothing but a wood fire where clay pottery was placed in with the burning coals to be fired. The American Indians made pottery in this same manner. The Romans actually developed a rather extensive ceramic industry including the making and firing of the following products:

Pottery	Dinnerware
Bricks	Tiles
Clay Conduits	Clay Pipes
Terra-cotta	Decorations for buildings

The Romans used at least two kinds of kilns. The standard pottery kiln would have a permanent floor and walls made of clay brick with a roof made of green wood coated with clay which was replaced after every firing. The Romans made bricks in field kilns which were nothing more than a large pile of unfired bricks with a tunnel in the bottom of the pile connected to vertical flue holes that went from the bottom tunnel to the top of the pile. A wood fire was built into the tunnel and kept burning for 4 to 6 weeks. The hot exhaust from the wood fire would flow through the vertical flues which heated the center of the pile enough to fire the brick in the core of the pile. The outside bricks, which acted as insulation, did not get fired. The pile was then allowed to cool and was torn down to get to the fired brick in the center.

The Roman kilns described above, were batch type which were also used by early Egyptians and Chinese civilizations, however, possibly as early as 500 A.D. the Chinese developed an ingenious kiln which recuperated heat in the same manner as continuous kilns do today. This Chinese kiln was built on a hillside with a series of steps and a side door located at each step. Pottery to be fired was placed on each step going up the hillside. A wood fire was built at the bottom of the hill and since the hillside kiln acted like a chimney, the exhaust gases from the fire would travel by natural buoyancy from step to step and exit at the top of the hill. The pottery was, therefore, preheated by the exhaust gases coming from the lower steps which greatly reduced the amount of wood required to fire the pottery. The fires would gradually be moved up the hill to complete the firing of all levels.

Kiln technology stayed much the same until the 1800's when coal became available as a much more efficient fuel for firing ceramics. The batch fired kilns evolved into the downdraft periodic kilns commonly called beehive kilns as shown in Illustration #26. This type of kiln was built with fire clay refractory brick, and had a dome which provided a permanent roof.

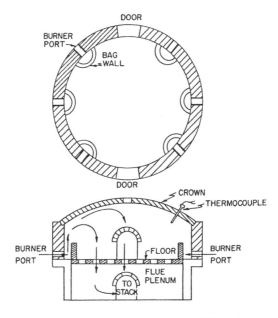

Illustration #26 – Beehive Kiln Functional Drawing.

The floor contained a series of flue openings which connected through an underground, refractory lined tunnel to a large brick stack. The stacks were often 30 meters high and provided the draft for up to four to five kilns. Illustration #26 shows the drawing of a beehive type periodic that has been converted to gas or oil fuel; however, the original beehive kilns were all coal fired so that a coal fire was built at each bag wall. In other words, with the kiln shown in the diagram, there are 6 bag walls so there would have been 6 separate coal fires. The hot gases from the coal fires would rise naturally to the roof by bouyancy, however, the draft from the chimney would pull the heated gases down through the load and through the floor to the stack.

Illustration #27 – Coal Fired Beehive Kiln Converted to Gas.

Illustration #27 shows a photo of a coal fired beehive kiln that has been converted to gas firing. This conversion was done and the photo was taken in 1950. Downdraft periodics could fire large loads (50,000 to 60,000 brick) however, they were inefficient from a fuel standpoint and very slow with regard to heating and cooling.

Also during the 1800's the chamber kiln was developed which was much more fuel efficient than the coal fired periodics. The chamber kilns were also coal fired, however, they used the principle of preheating the ware with exhaust gases similar to the old Chinese hillside kiln and the modern tunnel kiln. The chamber kiln was, in fact, the forerunner of the tunnel kiln.

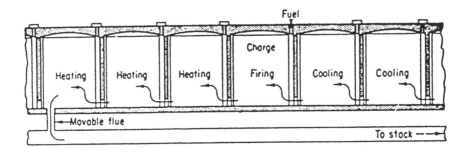

Illustration #28 – Chamber Kiln Functional Drawing.

Illustration #28 shows a functional drawing of a chamber kiln. The flue was movable from chamber to chamber so that each chamber became a preheat chamber, a firing chamber or a cooling chamber depending on where the flue was located. The diagram in Illustration #28 shows fuel which was granular coal coming in through the roof in the chamber which is presently the firing chamber. The coal is combusted with hot combustion air which is pulled through the cooling chambers to extract the heat from the hot ware. The hot exhaust gases are pulled out of the firing chamber and through the three preheating chambers to the movable flue. After all the ware in the firing chamber has matured, the fuel input is moved to the adjacent preheat chamber and the flue is moved from the previous first preheating chamber to a chamber that has just been filled with green ware.

EVOLUTION OF MODERN KILNS

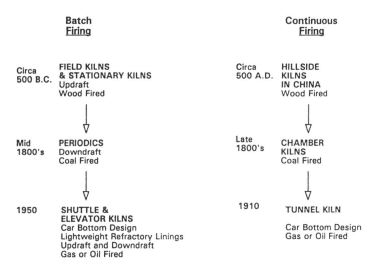

Illustration #29 – Evolution of Modern Kilns.

Illustration #29 shows the evolution of modern kilns. You will note that batch firing kilns and continuous kilns evolved independently into today's modern designs. Five major breakthroughs in the 20th Century are responsible for the evolution of today's kilns. These major breakthroughs are as follows:

1. The switch from coal to oil and then eventually to natural gas as fuel. These changes made automatic operation feasible.
2. The use of kiln cars in both tunnel kilns and shuttle kilns as a means to move the product through the kiln in the case of continuous firing and in and out of the kiln in the case of batch firing.
3. The development of lightweight kiln refractories including both insulating firebrick and refractory fiber. Lightweight refractories greatly improved the efficiency of batch fired kilns where the kiln walls have to be heated and cooled evey cycle. The use of lightweight kiln cars in tunnel kilns had the same effect of substantially improving fuel efficiency.

4. The development of high velocity burner systems capable of firing with varying amounts of excess air had the effect of increasing temperature uniformity and, therefore, reducing firing cycle time for most ceramics.
5. The development of electronic controls for temperature, atmosphere, and pressure which have now evolved into computer controls.

7. KILN SELECTION

7.1 Fuel vs Electric

When selecting a kiln, one of the first decisions that has to be made is to decide whether the kiln should be electric or fuel fired. This decision can best be made considering the advantages and disadvantages of each type of firing system.

THE ADVANTAGES OF ELECTRIC OVER FUEL FIRING ARE AS FOLLOWS:

A. Special atmospheres
B. Cleanliness
C. Lower capital cost in small sizes

A. *Special Atmospheres*

Since electric kilns do not have an exhaust system and are in essence, a closed environment, they are ideal for firing ceramics where a special atmosphere other than air is required. Products such as soft ferrites which are fired in a nitrogen atmosphere are usually fired in continuous electric kilns.

B. *Cleanliness*

In situations where cleanliness is very important, such as decorating dinnerware and art procelain, electric batch kilns or electric continuous belt kilns are commonly used. At decorating temperatures, the glazes are wet so that any airborne dirt will stick to the glaze and make a defect. Since there is no air flow in electric kilns, the airborne dust problem is eliminated.

C. *Lower Capital Costs in Small Sizes*

Small size electric kilns such as laboratory kilns, hobby kilns, and art pottery kilns can be made more inexpensively than the same size gas kilns. As a result, the overwhelming majority of small size kilns are electric.

THE ADVANTAGES OF FUEL FIRING OVER ELECTRIC ARE AS FOLLOWS:

A. Lower operating cost
B. Better temperature uniformity in large size kilns
C. Lower capital cost in large size kilns
D. Lower maintenance cost
E. Higher temperature capability

A. *Lower Operating Cost*

Heat generated from natural gas or oil costs considerably less than heat generated by electricity. This cost differential is so great that it even overcomes the extra heat loss that occurs in fuel fired furnaces through the exhaust system.

B. *Better Temperature Uniformity in Large Size Kilns*

Due to the convection heat transfer in fuel fired kilns, they can be made in large cross sections. Good temperature uniformity is achieved by circulation of the heating gases within the fuel fired kiln. In an electric kiln, there is no circulation and the heating is all done by radiation. As a result, electric kilns must be limited to small cross sections in order to get acceptable temperature uniformity.

Illustration #30 shows typical graphs for heating loads of ceramic ware by radiation and convection only. By contrast, high velocity convection can speed up heat flow to pieces obscured in the center of the load.

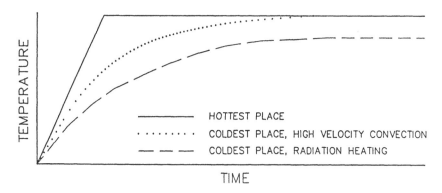

Illustration #30 – Heat Transfer to Coldest Piece with Radiation and Convection.

C. *Lower Capital Cost in Large Capacity Kilns*

Large capacity electric kilns cost considerably more to build because the cross sections must remain small and this makes the kilns very long in order to get capacity. A long narrow kiln costs more to build than a short, wide kiln. For example, in order to get a certain level of temperature uniformity, an electric tunnel kiln might be limited to a 600 mm wide setting width, whereas a fuel fired tunnel kiln could be 5 or 6 times wider than that. In this example, in order to get the same capacity, the electric kiln would have to be 5 or 6 times longer.

D. *Lower Maintenance*

Fuel fired kilns generally cost less to maintain than electric kilns because burners last a very long time whereas, heating elements must be periodically replaced. The higher the temperature, the more frequently the electric elements have to be replaced and the more costly they are.

E. *Higher Temperature Capability*

Fuel fired kilns with preheated combustion air can reach temperatures as high as 1850°C whereas 1650°C is pushing the practical limit for electric elements.

For the above stated reasons, the great majority of all ceramics are fired in fuel fired kilns. However, electric firing definitely has its place and is the preferred means of firing for some of the newer advanced ceramics which are either made in small quantities or require special atmospheres.

7.2   Batch vs Continuous Firing

The second major decision that must be made in selecting a kiln is to decide whether the kiln should be a batch fired kiln or a continuous firing kiln. In order to intelligently make this decision it is necessary to consider the advantages and disadvantages of both types of firing systems.

THE ADVANTAGES OF CONTINUOUS FIRING ARE:

A. Uses less fuel
B. Lower capital cost in large sizes
C. Easier to automate
D. Can fire faster cycles

A. *Continuous Kilns Use Less Fuel*

When comparing a modern continuous kiln with a modern batch kiln, the modern continuous kiln should use approximately one half of the fuel in firing the same product on the same cycle as the batch kiln. There are two reasons for this, the first being that in the continuous kiln; the walls are only heated one time and the kiln may run continuously for years; therefore, the heat storage in the walls does not have to be replaced every cycle as in batch kilns. However, the heat storage in the kiln car refractories is part of the daily fuel usage in a continuous kiln since the cars are heated and cooled as they move through the kiln.

The main reason for the difference in the fuel consumption between continuous and batch firing kilns is the recuperation which occurs in continuous kilns as the exhaust gases are moved counter-flow to the incoming product. This transfers heat into the incoming product that would normally be lost up the stack in a batch kiln. The temperature of the exhaust gases is proportionate to the amount of heat being lost up the stack. In a batch kiln, the temperature of the exhaust gases tracks the temperature inside the kiln so that in the beginning of the cycle the exhaust temperature will be approximately 100°C and by the end of the cycle the exhaust temperature will be equal to the soak temperature. Whereas in a continuous kiln, the exhaust gas temperature may be held at a constant 200°C to 250°C.

B. *Continuous Kilns are Lower Capital Cost for Large Capacity Kilns*
Large capacity kilns, such as the 300 and 400 ft. long tunnel kilns used to fire face brick, usually cost less to build than the equivalent capacity in shuttle kilns.

C. *Continuous Kilns are Easier to Automate for Loading and Unloading*
Continuous kilns are generally easier to automate than batch fired kilns especially conveyor type kilns such as roller hearths, belt kilns, and pusher slabs. Also, one-high tunnel kilns are in the same category. The reason that these type of kilns are easier to automate is the one-high setting does not require elaborate mechanical equipment to stack up the product and de-stack the product. The fact that continuous kilns are usually firing the same product at the same speed around the clock also lends itself to easier automation.

D. *Continuous Kilns can Fire Faster Cycles*
Although it is possible today to fire fiber lined batch kilns as fast as 8 to 10 hours, cold to cold, continuous kilns can be fired much faster. A one-high tunnel kiln with fiber kiln cars can fire as fast as 6 hours, cold to cold. Roller hearth kilns can fire as fast as 30 minutes, cold to cold, depending on the product.

THE ADVANTAGES OF BATCH FIRING ARE AS FOLLOWS:

A. Flexibility

B. Lower capital cost in smaller sizes

C. Can fire high setting heights
D. Can fire unusual firing cycle profiles

A. *Batch Kilns are More Flexible*

Flexibility is the biggest single advantage of batch firing kilns and is the reason why more batch kilns are built than continuous kilns. However, since continuous kilns are generally larger than batch kilns, the total volume of ceramics fired in continuous kilns is greater than the total volume fired in batch kilns. Flexibility means a batch kiln can fire three different products on three different firing cycles in the same week, which would not be practical to do in a continuous kiln. For example, in the grinding wheel industry they fire their small grinding wheels on a 48 hour cold to cold cycle, medium size wheels on an 84 hour cycle cold to cold, and their larger sizes on a 120 hour cold to cold cycle. The same batch firing kiln can fire all three of those cycles and the firing cycle mix can vary to suit market conditions.

Flexibility of firing capacity is another advantage of the batch fired kiln. For example, when introducing a new product, the kiln could be sized to fire the initial requirement for the product in one two-day cycle per week. As the requirement for the product grows, the kiln is simply fired twice a week and then three times a week until such time as a second kiln is needed. It is not efficient to use a continuous kiln in the same manner. For example, you would have to start out with an oversized continuous kiln running at a very slow speed. As the product demand increased, the kiln push rate would be gradually increased; however, the operating costs would be about as high as if you were running at capacity. By the time the kiln gets to capacity, it would then be time for another kiln. For this reason, most new products are first fired in a batch kiln and only after an established market is achieved is continuous firing added to batch firing capacity.

B. *Batch Kilns Have Lower Capital Cost in Smaller Sizes*

Generally speaking, there is a minimum size for a tunnel kiln below which it is less costly to build a batch kiln. This minimum size varies from product to product and is also dependent on firing cycle speed.

A typical example would be a 12 cu. meter batch fired kiln with a usable setting space of 2 meters wide × 4 meters long × 1.5 meters high. To load this same volume in a tunnel kiln, the tunnel kiln might have a setting width of 600 mm, a setting height of 600 mm and an overall length of 33 meters. Both kilns would hold 12 cu. meters of product and when firing the same cycle would fire the same quantity of product per day, however, the tunnel kiln would cost more to build.

C. *Batch Kilns can Fire High Setting Heights*

Products where a high setting height is required such as large clay pipes, large hollow electric insulators, and long ceramic tubes, are better fired in batch fired kilns. The reason for this is that height is the most difficult dimension to control from a temperature uniformity standpoint in a continuous kiln. For this reason, modern tunnel kilns are usually built with low setting heights. On the other hand, updraft batch fired kilns can be built where the heating gases flow up through and around the product to provide excellent temperature uniformity from top to bottom.

D. *Batch Kilns Can Fire Unusual Firing Cycle Profiles*

The typical firing cycle profile for a continuous kiln is more or less a straight line heat up to soak and a straight line cool down after soak, even though this can vary somewhat. Some ceramic products require a very severe deviation from a straight line such as a substantial hold period prior to the normal soak period. This is difficult to achieve in a continuous kiln but is quite easy in a batch kiln. An example of this is when firing silica coke oven blocks which must be heated very slowly up to 600°C because of the volume change which occurs with the alpha to beta quartz inversion. After the inversion, the heating rate can be much more rapid up to soak. The cooling curve is similar to the heatup with fast rates possible above the inversion temperature and very slow rates are required below 600°C. As a result, silica coke oven blocks can be fired much more efficiently in shuttle kilns than in tunnel kilns.

In summation, large constant volume production of the same product is usually best fired in continuous kilns. Most other situations are best fired in batch kilns. Sometimes the most efficient firing system is not either, just batch or just continuous but a combination of both. A good example of this would be where you have a high volume market which fluctuates. Rather than having one large tunnel kiln which is constantly being speeded up and slowed down, it would be better to build a tunnel kiln for 80% of the peak production requirement and have a shuttle kiln handling fluctuations between 100% and 80%

Based on the above mentioned advantages for continuous and batch firing, the following chart shows the preferred method for firing various products.

Products Usually Fired in Continuous Kilns

    Facebrick
    Roof Tile
    Sanitaryware – First Fire
    Dinnerware
    Ceramic Tile
    Ferrites
    Sparkplugs
    Small Electric Insulators
    Small Clay Pipe
    Standard Refractory Brick

Products Usually Fired in Batch Kilns

    Advanced Ceramics
    Abrasives
    Refractories other than standard brick
    Medium and large electric insulators
    Carbon Products
    Ceramic Colors and Pigments
    Large Clay Pipe
    Sanitaryware – Refire

8. KILN CONSTRUCTION

8.1 Refractory Linings

Traditionally, kiln refractory linings were made of dense refractory brick with steel I-beams called "buckstays" to support the exterior of the brick walls and to take the thrust of the sprung arch during heating. This old style of refractory construction is completely obsolete today due to the development of lightweight refractories. Up to one meter thick, dense refractory walls have now been replaced with much thinner insulating brick walls and fiber refractory walls. The lightweight refractory walls must be supported by a steel shell which is also used to support burners and piping on the outside of the kiln. Therefore, modern kiln refractory construction can be described as steel casings that support lightweight refractory linings varying in thickness between 200 mm and 550 mm depending on temperature.

Illustration 31 shows the prefabricated walls for a tunnel kiln with burners installed on the outside of the steel shell and the fiber refractory lining installed on the inside. This photograph was taken in the kiln builders factory after which this prefabricated kiln will be shipped to the jobsite.

Illustration 32 shows the installation of an insulating brick refractory wall. The 229 mm hot face insulating brick is backed up with 38 mm of block insulation. Block insulation has a very low conductivity factor (approximately 0.6) and as a result, greatly improves the insulating value of the wall without adding much cost.

Illustration 33 shows prefabricated tunnel kiln modules with a fiber refractory lining. These modules are assembled in the kiln builders factory

Illustration #31 -- Prefab Walls for Tunnel Kiln. Courtesy of Swindell Dressler.

Illustration #32 -- Insulating Firebrick Installation. Courtesy of Swindell Dressler

and will be shipped to the jobsite for final assembly. An advantage of this type of kiln construction is that the tunnel kiln can be moved at a later date if required.

Illustration 34 shows the installation of fiber refractory modules in a kiln wall. Fiber refractory modules are very quick to install using a special stud welding gun so that if the conditions are right for the use of standard fiber, the installed cost of the fiber lining will be less than the equivalent insulating brick lining.

Illustration 35 shows an updraft shuttle kiln lined with fiber above the hearth level and insulating fire brick below hearth level. There are places where lightweight insulating brick and fiber linings cannot be used such as in the hot zone of a high temperature tunnel kiln or in the preheat zone of a tunnel kiln where corrosive volatiles are present. In these cases, dense refractories are required.

Illustration #33 – Tunnel Kiln Modules with Fiber Lining. Courtesy of Swindell Dressler.

Illustration #34 – Fiber Module Installation. Courtesy of Thermal Ceramics Inc.

Illustration #35 – Shuttle Kiln Lined with Fiber. Courtesy of Swindell Dressler.

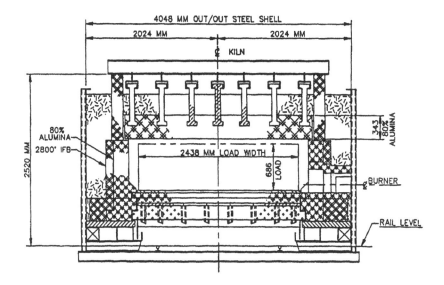

Illustration #36 – Tunnel Kiln with Dense Refractory Lining. Courtesy of Swindell Dressler.

Illustration #36 shows a dense refractory lining in a high temperature tunnel kiln. The dense refractories are, of course, backed up with lighter weight refractories to minimize the thickness of the wall.

Illustration #37 shows the heat loss and heat storage for typical kiln walls. All three examples were designed for a similar heat loss per square meter through the wall; however, the heat storage in a square meter area of each type wall is radically different. This shows why batch kilns are almost never lined with dense refractory walls. With continuous kilns where heat storage is not a factor, the selection criteria is the cost to build and the projected life of the refractory.

Refractory linings for kiln cars are quite different than the refractory linings in the walls and roof of kilns because the considerations are different. Since kiln cars are periodically heated and cooled regardless of whether they are used in batch or continuous kilns, the heat storage in the car refractory is very important. Since heat storage is directly proportional to the weight of the refractories, it is desirable to have the lightest weight refractories possible. However, the kiln car also must support the load and only dense refractories have the load bearing qualities required. The answer to these two problems is that the ideal design for a kiln car is to have only enough dense refractories to support the load with the balance of the refractories in the car being as light as possible.

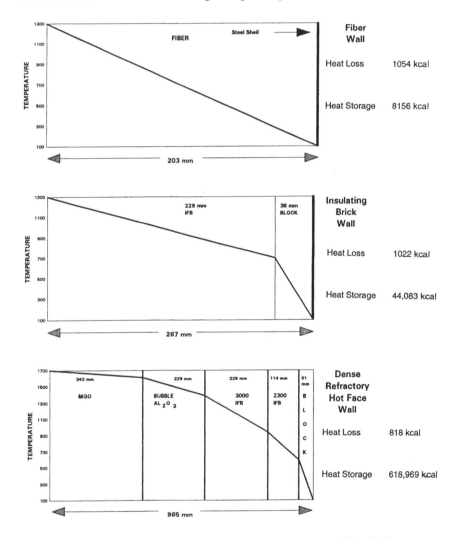

Illustration #37 – Heat Loss and Heat Storage in Typical Kiln Walls.

Illustration #38 – Kiln Car with Downdraft Flue for Heavy Load. Courtesy of Swindell Dressler.

Illustration #38 shows a kiln car designed for 1500°C use and a heavy load. In this case, the top of the car is made of alumina plates to carry the load. All the rest of the refractories are insulating bricks and block insulation. The kiln car shown in this illustration happens to be for a batch fired kiln with a downdraft exhaust system, as can be seen by the flue in the center of the car.

The type of kiln car used in updraft shuttle kilns and continuous push tunnel kilns, where the burners underfire the load, is shown in Illustration #39. The refractory lining on these kiln cars is not as heavy as it looks. The blocks above the girders which support the load are formed like grating so that the heating gases can flow up through them. The girders which support the grate blocks have less weight than solid piers with a rectangular cross section. The blocks below the girders are hollow core blocks. The refractories below them, except on the perimeter of the cars, are insulating brick.

Illustration #39 — Kiln Car for Under-Firing. Courtesy of Swindell Dressler.

The lighter the product load, the lighter weight the kiln car refractory lining can be. Illustration 40 shows an old style heavy kiln car for a sanitaryware tunnel kiln. Illustration 41 shows the same kiln car rebuilt as a light weight car. The light weight car uses fiber as heat insulation and cordierite hollow posts which support hollow recrystalized silicon carbide structural beams to support the load. The light weight car weighs only 10% as much as the heavy car. The light weight car not only saves fuel on every cycle but also can be fired faster and reduces maintenance cost.

Illustration #40 – Old Design Heavy Kiln Car. Courtesy of Swindell Dressler.

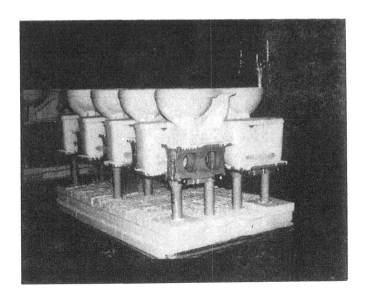

Illustration #41 – New Design Light Weight Kiln Car. Courtesy of Swindell Dressler.

The most economical type of refractory lining in fuel fired kilns, assuming there is no corrosive environment, is as follows:

MOST ECONOMICAL REFRACTORY LINING FOR FUEL FIRED KILNS (assuming no corrosive environment)

KILN MAXIMUM FIRING TEMPERATURE	BATCH KILN	CONTINOUUS KILN KILN
Up to 1300°C	Fiber Modules 1400°C grade	Fiber Modules 1400°C
1300° to 1550°C	Insulating Firebrick	Insulating Firebrick
1550° to 1800°C	Bubble Alumina Brick backed up with insulating firebrick	Dense Refractory backed up with insulating firebrick

The previous chart should only be used as a guide in selecting the appropriate refractory lining since there are exceptions which would warrant a different type of lining. For example, you will note that fiber modules are recommended only up to 1300°C. Fiber modules are manufactured for temperatures as high as 1650°C, however, they are usually not cost effective when compared to insulating firebrick at temperatures above 1300°C. However, if you were building a batch fired kiln with a very fast firing cycle, such as 12 hours cold to cold for temperatures up to 1500°C, you probably could justify the use of the high temperature fiber modules based on fuel savings. Another example of an exception would be in a tunnel kiln for use at 1700°C where dense alumina would normally be less expensive and last longer than bubble alumina, however, if the kiln might have to be moved someday, modular construction with a lightweight lining would be preferable.

The above chart only refers to fuel fired kilns. Due to the fact that heat storage is much more important in electric kilns, fiber is normally recommended at all temperatures for electric batch kilns. Most electric continuous kilns also use fiber with the exception of the pusher slab kilns used for firing soft ferrites. In this case, the open pore structure of fiber is detrimental to maintaining the high purity atmosphere required.

8.2 Kiln Geometry

Kiln geometry is one of the most important considerations in designing a kiln because it has a very substantial effect on temperature uniformity, capacity, and the cost to build the kiln. One major factor that affects kiln geometry is the method of heat transfer, radiation vs. convection. Electric kilns are obviously designed for radiant heat transfer which dictates a small dimension in the direction that the heat is flowing; for example if the elements are on the sides, that would dictate a narrow setting width. If the elements are on the top, that would dictate small setting height. The reason for this is that radiation heats only what it sees which would be the outside layer of ceramic product closest to the elements. The heat must then be conducted across the ceramic (since most ceramics are insulators this is relatively slow). The heat is then radiated across the air space from the first row of ceramic product to the second row; after which the heat is then conducted through the second row of ceramic product and so on from row to row.

Fuel Fired systems use a combination of convection and radiation heat transfer with convection being the most important at lower temperatures and radiation becoming more important at higher temperatures. Regardless of temperature, the controlled flow of heating gases through the kiln can be used to create uniform temperatures even in very large cross section kilns. Modern downdraft and updraft fuel fired kilns are designed to distribute the heating gases uniformly through the entire cross section. Setting widths of 6 to 8 meters are possible with good temperature uniformity in either batch or continuous kilns. Setting heights of as much as 5 meters in batch kilns are feasible with good control.

Unlike batch kilns, continuous kilns have a limitation in setting height. Tunnel kilns function like a chimney laid over on its side in a horizontal position so that the hotter gases flow down the top of the tunnel, this creates a substantial temperature gradient from top to bottom. For years, kiln builders tried to overcome this problem with fans used to pull heat off the top and blow it into the bottom with marginal results. With the advent of jet burners which were capable of firing across a wide kiln cross section, it became practical to build tunnel kilns with low setting heights and wide setting widths. The low setting height provides good temperature uniformity top to bottom and the wide setting width recovered the capacity that was lost when the setting height was lowered.

The following chart shows some rules of thumb used by kiln designers in determining kiln geometry. There obviously are a number of exceptions to these rules, however, they are at least a good starting point when planning a kiln design.

RULE OF THUMB
LIMITING SETTING DIMENSIONS

Electric Batch Side Elements	600 mm width maximum
Electric Continuous Top Elements	300 mm height maximum
Fuel Fired Batch	No reasonable limits (8 meters widths are possible) (5 meters heights are possible)
Fuel Fired Tunnel Kiln	1500 mm height maximum
Fuel Fired Roller Hearth	One piece high

9. ELECTRIC KILNS

9.1 Types of Elements

There are five types of elements commonly used in ceramic kilns. Two of these are metallic, two are ceramic and one a cermet. A discussion of each type of element is as follows:

A. *Metallic Elements*

The two types of metallic elements are nickel/chrome (NiCr) and iron-chromium-aluminum (Fe-Cr-Al). The metallic elements are the least expensive, however, they have the lowest use temperatures. NI-Cr is limited to use at 1100°C whereas the iron-chromiun aluminum alloys (Kanthal A-l) can be used at kiln temperatures up to 1300°C. Nichrome, although limited to the lower temperatures, has very good hot strength so the elements can be self supporting whereas the Fe-Cr-Al elements with poorer hot strength must be supported usually by ceramic tubes.

Illustration 42 shows various installations of Fe-Cr-Al element wire which is wrapped around a mullite tube for support. This element system is the most inexpensive system for use up to 1300°C and also offers very long life.

Illustration 43 shows various installations using nichrome electric elements in ribbon form. These nichrome elements are 80% nickel and 20% chrome which is the most common composition. Both types of metallic elements have the advantage that their electrical resistance stays constant with time so that as the elements age it is not necessary to make compensations for changing resistance. Also, the metallic

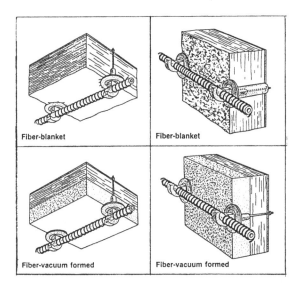

Illustration #42 – Metallic Elements – Wire (Fe-Cr-AL). Courtesy of Kanthal Corporation.

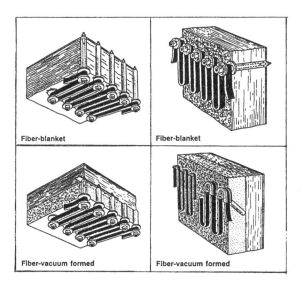

Illustration #43 – Metallic Elements – Ribbon. Courtesy of Kanthal Corporation.

elements have more or less constant resistance at all temperatures. As a result of the fact that the resistances are constant, inexpensive on/off controls can be used with metallic elements.

The Nichrome element forms a chromiumoxide layer when heated in the presence of air. The oxide layer is relatively thick, greenish in color and has a propensity to flake off during cycling. This flaking exposes the base material to further oxidation which eventually leads to the element failure. This flaking can also lead to product contamination but the element should be located in such a position that the oxide will not land on the product.

The Fe-Cr-Al alloys form an oxide which is mainly alumina (Al_2O_3). The alumina coating is very thin and is less likely to flake off or cause contamination. As the elements are cycled, small cracks may develop in the oxide coating which leads to aluminum depletion in the base metal. Generally, the rate of depletion is less than for nichrome materials which translates into longer life.

B. *Silicon Carbide Elements*

Silicon carbide elements are the least expensive heating elements for the temperature ranges between 1300 and 1500°C. Silicon carbide elements cost about half as much as the higher temperature elements and they also have the advantage that they can be mounted horizontally and are self supporting. Silicon carbide elements are generally made in rod form and have a hot center zone and two cold ends. The cold ends are impregnated with silicon metal so that they offer very low resistance and minimize power losses. Illustration 44 shows various sizes of straight and multi leg silicon carbide elements. Silicon carbide elements can take a much higher watt loading per square centimeter than metallic elements and, therefore, fewer elements are required to obtain the same heat input.

Silicon carbide elements are manufactured from grains of silicon carbide which are bonded together in a sintering process. Sintering causes bridges between the grains which provides a means for curren flow through the element. Over a period of time, the SiC bridges between the grains will slowly oxidize to silica (SiO_2) which is a poor conductor of electricity. As a result, the resistance of the element increases with time and this process is called aging. Over the lifetime of the silicon carbide elements the resistance will generally increase by a factor of 4. Silicon carbide also exhibits a changing resistance with temperature. The resistance is fairly high at room temperature but falls to a minimum value at about 800°C. At element temperatures above 800°C, resistivity increases with rising temperatures. Due to the characteristics of aging and resistance change, silicon carbide elements cannot use inexpensive on/off controls but must use silicon controlled rectifiers (SCR control). SCR control is more expensive than on/off control but can handle the increased voltage as the elements age and also can limit the current during the negative portion of the resistance curve.

Illustration #44 – Silicon Carbide Elements. Courtesy of Kanthal Corporation.

C. *Molybdenum Disilicide*

Molybdenum disilicide is a cermet made by a powder metallurgical process in which the base material has certain ceramic and metallic components. When this element material was first developed, it was limited to a maximum element temperature of 1700°C which would translate to a kiln maximum temperature of 1600°C. However, today higher grades are available such as Kanthal Super 1900 which permits element temperatures of 1900°C and kiln temperatures of 1800°C.

Molybdenum disilicide exhibits a very large increase in resistance from room temperature to operating temperature. This increase can be in the range of 10 to 14 times. Unlike silicon carbide, the material does not age and the resistance remains relatively stable over the life of the element. Due to the change in resistance with temperature, these type elements require SCR control instead of simple on/off control for the power.

In the presence of oxygen, molybdenum disilicide elements will form a protective oxide of silicon dioxide or quartz glass on the surface. Since molybdenum disilicide is very dense with a porosity of less than 1% the oxidation occurs only on the surface of the element. Further, the oxide which is usually very thin adheres tightly and rarely flakes off. The glass coating on the surface of the element, therefore, offers excellent protection against further oxidation extending the life of the element often to several years.

Molybdenum disilicide elements are always made in the shape of a hairpin, mounted vertically with open ends at the top. If molybdenum disilicide elements are mounted horizontally, they must be fully supported by laying on a refractory plate because the material itself softens at 1200°C.

Illustration 45 shows a small box kiln with molybdenum disilicide elements mounted vertically on the side walls.

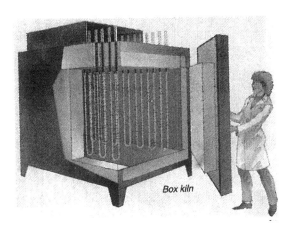

Box kiln

Illustration #45 – Molybdenum Disilicide Elements. Courtesy of Kanthal Corporation.

D. *Zirconia Elements*

Zirconia elements are the only elements that can be used in an air atmosphere at temperatures higher than molybdenum disilicide elements. Zirconia elements have only been used in laboratory size kilns because they are only available in small sizes and are very expensive. Also, these elements must be preheated to 1000°C before conduction even starts. Zirconia elements can be used at kiln temperatues up to 2000°C and SCR controls are required.

TYPES OF ELECTRIC ELEMENTS
(For Use in Air)

	Nickel-Chrome NiCR	Iron-Chrome -Aluminum Fe-CR-AL	Silicon Carbide SiC	Molybdenum Disilicide $MOSi_2$	Zirconia ZrO_2
Max. Element Temp. (Air)	1200°C	1400°C	1650°C	1900°C	2200°C
Max. Use Temp. (Air)	1100°C	1300°C	1550°C	1800°C	2000°C
Max. Use Temp. (N_2)	1120°C	1100°C	1350°C	1650°C	
Element Material Class	Metallic	Metallic	Ceramic	Cermet	Ceramic
Electrical Resistance vs. Time (Aging)	Constant	Constant	Increases 4 times	Constant	Constant
vs. Temperature	Constant	Constant	Decreases to 800°C then slowly increases	Increases 10–14 times	Conduction starts at 1000°C
Type Control Required	On/Off	On/Off	SCR	SCR	SCR
Major Limitation	Limited to 1100°C	Poor hot strength Needs element support	Elements age	High cost	Must be preheated to 1000°C

Illustration #46 – Types of Electric Elements.

E. *Summary*

Illustration 46 shows a chart entitled "Types of Electric Elements". This chart shows the major differences between the various types of electric elements and can be used as a guide for selecting the appropriate element system for your kiln.

9.2   Batch Electric Kilns

The three main applications of Batch electric kilns are:

1. laboratory kilns because they are the least expensive small kilns and easy to operate
2. decorating kilns for glazed ceramics because of cleanliness
3. kilns where an atmosphere other than air is required

THE TYPES OF ELECTRIC BATCH KILNS ARE AS FOLLOWS:

A. *Box Kilns*

Illustration 47 shows two different size box kilns. The box kiln is the most inexpensive way to build a small kiln. This kiln has a solid bottom as compared to other types of batch kilns which have a car bottom. Box kilns shown in Illustration 47 use metallic elements;

Illustration #47 – Electric Box Kilns. Courtesy of Unique/Pereny.

however, box kilns can be built with any type of element. Box kilns usually have elements on the side and underneath the hearth and are lined with either fiber or lightweight insulating brick.

B. *Shuttle Kilns*

Illustration 48 shows a typical electric shuttle kiln used for firing art pottery. Electric shuttle kilns are the most popular type of kiln for art pottery fired in an oxidizing atmosphere because of the clean environment with electric heating and the lower price compared to gas kilns.

Illustration #48 – Electric Shuttle. Courtesy of Unique/Pereny.

9.3 Continuous Electric Kilns

The main applications for continuous electric kilns are:

1. decorating kilns for glazed ceramics due to cleanliness
2. kilns where an atmosphere other than air is required

There are four different types of electric continuous kilns, namely: the tunnel kiln, belt kiln, roller hearth kiln, and the pusher slab kiln. The electric tunnel kiln today has been pretty much replaced by the other three types of continuous electric kilns. The pros and cons of each design are discussed as follows:

A. *Electric Belt Kilns*

Electric belt kilns are the most popular type of kiln today for high volume decorating of glazed ceramics. The electric belt kiln has the advantage of cleanliness plus, the belt eliminates the need for kiln furniture. Finally, belt kilns are capable of very fast decorating cycles; one hour or less. The major limitation is that the maximum use temperature for the stainless steel belts are approximately 950°C. Belt kilns are also used in ceraming operations for recrystalized glass products. Illustration 49 shows an electric belt kiln for decoration.

Illustration #49 – Electric Belt Kiln.

The recent requirement for lead-free glazes on dinnerware has elevated some decorating temperatures beyond the capability of stainless steel belts. As a result, roller hearth kilns will take over some of the dinnerware decorating operations that were formally performed with belt kilns.

B. *Electric Roller Hearth Kilns*

Roller hearth kilns generally cost more to build than belt kilns; however, they can be used for much higher temperatures. Therefore, an electric roller hearth kiln fits in where the temperature is too high for a belt kiln and also where very fast cycles are required which would be difficult to accomplish in a tunnel kiln.

C. *Electric Pusher Slab*

Pusher slab kilns are ideally suited for firing ceramics in a protected atmosphere. The reason for this is that the plates carrying the product can be pushed through a kiln which is sealed on all four sides. The steel shell of the pusher plate kiln is usually made to be gas tight.

D. *Electric Tunnel Kilns*

Electric tunnel kilns have their heating elements on the sidewalls and, therefore, must be very narrow in cross section. The narrow cross section is due to the fact that heat is transferred only by radiation. If the tunnel kiln were to have a wide cross section, it would have a cold middle; therefore, the electric tunnel kiln must be made very long in order to increase capacity. Since length is the most costly dimension of a kiln to build, electric tunnel kilns are expensive.

Another reason why electric tunnel kilns have gone out of favor is that both electric belt kilns and electric roller hearth kilns which have the heating elements above and below the belt or rolls are capable of more uniform temperatures and faster cycles. Lastly, in cases where a protective atmosphere is required, the electric pusher slab is a superior design to the electric tunnel kiln. With a pusher slab, all four sides of the kiln can be sealed, whereas with the tunnel kiln, the gap between the moving car train and the kiln walls is very difficult to seal.

Illustration #50 – Electric Pusher Plate – For Advanced Ceramics. Courtesy of Unique/Pereny.

10. FUELS AND FUEL EFFICIENCY

Over 90% of all ceramic products are fired in fuel fired kilns as was explained in Section 7.1 entitled, "Fuel vs. Electric". To reiterate, the advantages of fuel firing over electric are as follows:

1. Lower operating costs
2. Better temperature uniformity in large size kilns
3. Lower Capital cost in large size kilns
4. Lower maintenance cost
5. Higher temperature capability

10.1 Fuels

A. Natural gas
B. LPG (usually propane)
C. Light oil
D. Heavy oil (No.6)
E. Producer gas
F. Coal (powdered)

The pros and cons of each fuel are as follows:

A. *Natural Gas*

Natural gas is the preferred fuel for firing ceramics if available at a competitive price. Natural gas is clean, has a high calorific value and natural gas firing systems are less expensive than firing systems for other fuels. Another advantage of natural gas is that natural gas firing systems usually have less maintenance than firing systems for other fuels. Natural gas is usually delivered to the ceramic plant by the gas utility company and, therefore, fuel storage facilities are not normally required at the plant.

B. *LPG* (*liquid petroleum gas*)

If natural gas is not available, the next most desirable fuel for a ceramic kiln is liquid petroleum gas (LPG). LPG is clean burning, has a high calorific value and uses the same type of firing equipment as natural gas. However, an LPG kiln system requires a fuel storage facility at the ceramic plant. The fuel storage facility includes tanks to store the liquid petroleum gas under pressure as well as equipment to control the gasification of liquid fuel and heating equipment to prevent the lines from freezing. LPG is an ideal standby fuel for a natural gas installation, especially if the LPG is blended with air so that it equals the calorific value of natural gas per cubic meter of fuel. With this system, a kiln can be switched from natural gas to aerated LPG gas with very little adjustment to the kiln.

C. *Light Oil*

Light oil is usually defined as distillate #1 which is kerosene or distillate #2 which is diesel oil. Of these two fuels, diesel oil is preferred because diesel oil has lubricating qualities and kerosene does not. The advantage of the lubricating qualities of the fuel oil is that pumps, nozzles and valves do not wear out nearly as quickly as they would with kerosene. Light oil is the cleanest burning and the easiest to handle of the oil fuels, however, the oil firing equipment is more expensive and subject to greater maintenance than gas firing equipment.

For oil to burn, it essentially must be gasified in the burner which is done by atomization. Atomization can be accomplished by two methods; the first is spraying high pressure oil through a nozzle. In the high pressure oil method, substantial electric power is required to operate the high pressure pump. The second method of atomization uses compressed air which flows through the nozzle with the oil. Here again, electric energy is required to run the air compressor.

D. *Heavy Oil* (*No. 6*)

Heavy oil, or No. 6 oil, is a difficult fuel to use in a ceramic kiln; normally it would only be used if natural gas, LPG and light oil were not available or are economically unfeasible. Heavy oil is like tar and at room temperature you can hold a piece of it in your hand. In order to make heavy oil flow through pipes, it must be heated to approximately 60°C and then further heated to 110°C to obtain good atomization. The fuel pipes must be insulated and wrapped with an electric wire heating element to maintain the temperature in the pipe so the oil will flow. When heavy oil heating systems are turned off, the fuel lines must be flushed with compressed air to prevent the fuel from hardening inside the fuel lines. Heavy oil is not a clean burning fuel and often has impurities such as sulphur.

Another problem with heavy oil is that it may have some water content which can cause burners to go out. A method of handling this problem is to centrifuge the heavy oil before pumping it to the kiln. Heavy oil firing systems do not normally have a good turndown between minimum fire and maximum fire. As a result of this, it is easier to utilize heavy oil in a continuous kiln where high turndowns are not normally required compared to a batch fired kiln where higher burner turndowns are required.

E. *Producer Gas*

Producer gas, which is manufactured from coal, is another one of the fuels which would not be used in a ceramic kiln unless the premium fuels such as natural gas, LPG and light oil were not available. The disadvantage with producer gas is that it has a very low calorific value. As a result, the gas/air ratio can be as low as 2:1 which compares to natural gas which is 10.1. The problem of low gas/air ratio is that it requires that the gas lines are almost as large as the air lines and the burners themselves tend to get quite large. Because of the large gas lines, all the gas controlling hardware, such a regulators and valves, are much larger and more expensive. Also, since as much a five times the volume of producer gas needs to be heated to equal one volume of natural gas, the resultant flame temperature is much lower. Therefore, with producer gas it is difficult to achieve kiln temperatures much in excess of 1400°C.

F. *Coal*

Coal is very rarely used today as fuel for firing ceramics because it is so difficult to handle compared to liquid and gaseous fuels. Coal was commonly used as the fuel in the beehive, downdraft type periodics. Also many top fired index push tunnel kilns were built using coal as a fuel for firing common brick and face brick. The coal firing tunnel kilns have a very elaborate system for delivering powdered or very fine grained coal to a series of dispensers built into the roof of the tunnel kiln which spray in timed spurts of powdered or granular coal.

Comparative data (by weight) for some typical fuels

| | Heating value | | | | Gross Btu per scf | Wt air req'd per unit wt fuel | Weight of combustion products per wt of fuel | | | | Ultimate vol % CO₂ in dry flue gas |
| | Btu/lb (and Btu/gal) | | kcal/kg (and kcal/t) | | | | (and ft³/gal) | | | | |
	Gross	Net	Gross	Net	air[10]	(and scf/gal)	CO₂	H₂O	N₂	Total	
Blast furnace gas	1 179	1 079	665	599	135.3	0.57	0.58	0.01	1.08	1.67	25.5
Coke oven gas	18 595	16 634	10 331	9 242	104.4	13.63	1.51	1.81	8.61	11.93	10.8
Producer gas[1]	2 614	2 459	1 452	1 366	129.2	1.55	0.61	0.15	1.72	2.48	18.4
Natural gas[2]	21 830	19 695	12 129	10 943	106.1	15.73	2.55	2.03	12.17	16.75	11.7
Propane, natural	21 573	19 886	11 986	11 049	107.5	15.35	3.01	1.62	12.01	16.64	13.8
	(91 500)	(84 345)	(6094)	(5617)		(850.8)	(108.11)	(144.39)	(682.06)	(934.57)	
Butane, refinery	20 810	19 183	11 562	10 658	106.1	15.00	3.04	1.53	11.82	16.39	14.3
	(102 600)	(94 578)	(6833)	(6299)		(949.0)	(124.27)	(146.92)	(747.18)	(1018.4)	
Methanol	9 700	8 400	5 389	4 667	106.4	6.47	1.38	1.13	4.97	7.48	15.0
	(64 150)	(55 550)	(4272)	(3700)		(559.5)	(78.4)	(156.8)	(445.3)	(681)	
Gasoline, motor	20 190	18 790	11 218	10 440	104.6	14.80	3.14	1.30	11.36	15.80	15.0
	(123 361)	(114 807)	(8216)	(7646)		(1183)	(165.1)	(166.8)	(940.3)	(1272)	
#1 Distillate oil	19 423	18 211	10 791	10 118	102.1	14.55	3.17	1.20	11.10	15.48	15.4
	(131 890)	(123 650)	(8784)	(8235)		(1292)	(185.7)	(171.0)	(1020)	(1377)	
#2 Distillate oil	18 993	17 855	10 553	9 920	101.2	14.35	3.20	1.12	10.95	15.27	15.7
	(137 080)	(128 869)	(9130)	(8583)		(1354)	(199.1)	(170.6)	(1070)	(1440)	
#4 Fuel oil	18 844	17 790	10 470	9 884	103.0	13.99	3.16	1.04	10.68	14.92	15.8
	(143 010)	(135 013)	(9524)	(8992)		(1388)	(206.7)	(166.1)	(1097)	(1472)	
#5 Residual oil	18 909	17 929	10 506	9 961	104.2	13.88	3.24	0.97	10.59	14.81	16.3
	(149 960)	(142 190)	(9987)	(9470)		(1439)	(221.0)	(161.4)	(1137)	(1520)	
#6 Residual oil	18 126	17 277	10 071	9 599	103.2	13.44	3.25	0.84	10.25	14.36	16.7
	(153 120)	(145 947)	(10 198)	(9720)		(1484)	(236.4)	(149.0)	(1172)	(1558)	
Wood, non-resinous	6 300		3 500		98.4	4.90	1.39	0.65	3.47	5.51	20.3
Coal, bituminous	14 030		7 795		99.3	10.81	2.94	0.49	8.26	11.71	18.5
Coal, anthracite	12 680		7 045		97.8	9.92	2.96	0.22	7.58	10.78	19.9
Coke	12 690		7 051		96.2	10.09	3.12	0.07	7.73	10.94	20.4

Illustration #51 – Comparative Data by Weight for typical fuels. Courtesy of Combustion Handbook by North American Mfg. Co.

Note: An enlarged version of illustration 51 is shown in the Appendix on p. 212.

10.2 Fuel Efficiency

A. *Available Heat*

Fuel efficiency in the kiln has to do with available heat and where the heat goes. Available heat by definition is as follows:

The gross quantity of heat released within a combustion chamber minus both the dry flue gas loss and the moisture loss. It represents the quantity of heat remaining for useful purposes to heat the load and balance losses to walls, openings and conveyors.

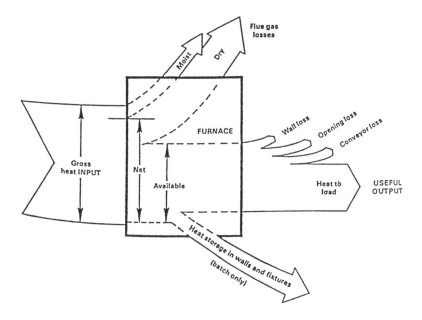

Illustration #52 – Sankey Diagram for a Furnace Heat Balance. Courtesy of Combustion Handbook by North American Mfg. Co.

Available heat can probably best be explained with a Sankey diagram, as shown in Illustration 52. With this diagram, the dark rectangle represents the kiln or furnace and the gross heat input from the fuel is shown coming into the kiln from the left. The flue gas losses, both dry and moist, are shown exiting from the top of the kiln.

The difference between the flue gas losses and the gross heat input is the available heat.

In batch fired kilns, some of the available heat is used as heat storage in the walls; however, in continuous kilns where the walls stay hot and do not have to be heated on every cycle, heat storage would not figure into the heat balance.

The balance of the available heat goes to heat loss through the walls, heat loss through various openings in the kiln, plus conveyor loss which includes heat storage in kiln cars or belts. Finally, after all of these losses, the heat that is left can be used to heat the load.

As can be seen from the Sankey diagram, the actual heat available to heat the load or the useful output, is usually less than half the total input to the kiln. In cases where the load includes kiln furniture, the kiln furniture is part of the load so that if the ratio between kiln furniture and actual product was one to one, the percentage of total heat input to the kiln that goes into the product would usually be less than 20%.

B. *Excess Air*

Available heat varies a small amount depending on the fuel being used such as natural gas, light oil or heavy oil; however, available heat varies greatly based on the amount of excess air being used. Illustration 53 shows the effect of excess air on available heat. The top line on this chart shows the available heat of natural gas when using 0% excess air. The subsequent lines on the chart show the available heat with varying amounts of excess air such as 25%, 50%, 200%, etc. The horizontal axis on the chart is the flue gas temperature at varying amounts of excess air. The 0% excess air temperature being the theoretical flame temperature. For example, increasing from 0% excess air to 100% excess air lowers the flame temperature from 1900°C to 1200°C. Also, when using extremely high percentages of excess air such as 1400%, the flame temperature or mixture temperature can be as low as 200°C which makes it possible to control kilns in the low temperature ranges.

Looking at the vertical axis of this chart, which shows percentage of available heat, it can be seen that the available heat with 0% excess air varies from a maximum of 84% at low kiln temperatures, all the way down to 0% as kiln temperatures approach theoretical flame temperature. For example, when using 200% excess air the available heat at 300°C is a little over 60%, however, the available heat goes to 0 at around 860°C, because at this point, the burner jet temperature is 860°C and no more heat can be transferred.

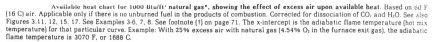

Available heat chart for 1000 Btu/ft³ natural gas*, showing the **effect of excess air upon available heat.** Based on 60 F (16 C) air. Applicable only if there is no unburned fuel in the products of combustion. Corrected for dissociation of CO_2 and H_2O. See also Figures 3.11, 12, 15, 17. See Examples 3-6, 7, 8. See footnote (†) on page 71. The x-intercept is the adiabatic flame temperature (hot mix temperature) for that particular curve. Example: With 25% excess air with natural gas (4.54% O_2 in the furnace exit gas), the adiabatic flame temperature is 3070 F, or 1688 C.

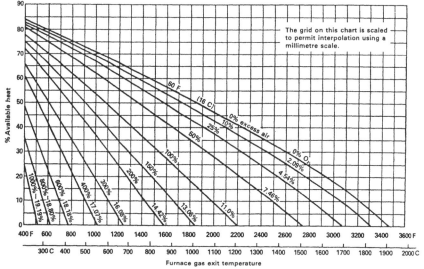

Illustration #53 – Effect of Excess Air on Available Heat. Courtesy of Combustion Handbook by North American Mfg. Co.

Note: An enlarged version of illustration 53 is shown in the Appendix on p. 213.

C. *Preheated Combustion Air*

Preheated combustion air is used for two reasons, First, it saves fuel and second, it elevates the flame temperature making it possible to fire to higher temperatures than would be possible without preheating the combustion air. Illustration 54 shows the effect of preheated combustion air on available heat and flame temperature. On this chart, the curve in the middle labeled 0% excess air, represents room temperature combustion air whereas the curves above that which are labeled 316°C, 427°C, 649°C, etc. represent varying temperature combustion air available heat curves with 10% excess air included. The 10% excess air curve with room temperature air indicates the flame temperature is approximately 1870°C. Whereas the available heat curve for 427°C preheated combustion air shows a flame temperature of approximately 2000°C. This flame temperature elevation is extremely important for firing ceramics above 1650°C. At these temperatures, without preheated air, the available heat equals the losses so effective heating stops.

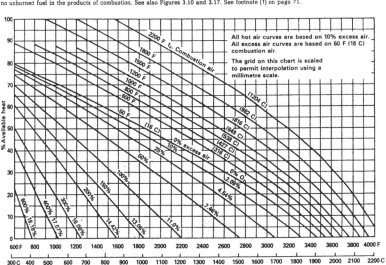

Available heat for 1000 Btu/ft³ natural gas with preheated combustion air at 10% excess air. Applicable only if there is no unburned fuel in the products of combustion. See also Figures 3.10 and 3.17. See footnote (†) on page 71.

Illustration #54 – Effect of Preheated Combustion Air on Available Heat and Flame Temperature. Courtesy of Combustion Handbook by North American Mfg. Co.
Note: An enlarged version of illustration 54 is shown in the Appendix on p. 214.

The other reason for using preheated air is fuel savings. Illustration 55 shows the fuel savings resulting from use of preheated air with natural gas and 10% excess air. In order to use this chart to evaluate

Fuel savings resulting from use of preheated air with natural gas and 10% excess air. These figures are for evaluating a proposed change to preheated air—not for determining system capacity. Read captions for Tables 3.16b and c.

% Fuel saved with natural gas, 10% XSAir	t_1, Combustion air temperature, F													
	600	700	800	900	1000	1100	1200	1300	1400	1500	1600	1800	2000	2200
1000	13.4	15.5	17.6	19.6	—	—	—	—	—	—	—	—	—	—
1100	13.8	16.0	18.2	20.2	22.2	—	—	—	—	—	—	—	—	—
1200	14.3	16.6	18.7	20.9	22.9	24.8	—	—	—	—	—	—	—	—
1300	14.8	17.1	19.4	21.5	23.6	25.6	27.5	—	—	—	—	—	—	—
1400	15.3	17.8	20.1	22.3	24.4	26.4	28.4	30.2	—	—	—	—	—	—
1500	16.0	18.5	20.8	23.1	25.3	27.3	29.3	31.2	33.0	—	—	—	—	—
1600	16.6	19.2	21.6	24.0	26.2	28.3	30.3	32.2	34.1	35.8	—	—	—	—
1700	17.4	20.0	22.5	24.9	27.2	29.4	31.4	33.4	35.3	37.0	38.7	—	—	—
1800	18.2	20.9	23.5	26.0	28.3	30.6	32.7	34.6	36.5	38.3	40.1	—	—	—
1900	19.1	21.9	24.6	27.1	29.6	31.8	34.0	36.0	37.9	39.7	41.5	44.7	—	—
2000	20.1	23.0	25.8	28.4	30.9	33.2	35.4	37.5	39.4	41.3	43.0	46.3	—	—
2100	21.2	24.3	27.2	29.9	32.4	34.8	37.0	39.1	41.1	43.0	44.7	48.0	51.0	—
2200	22.5	25.7	28.7	31.5	34.1	36.5	38.8	40.9	42.9	44.8	46.6	49.9	52.8	—
2300	24.0	27.3	30.4	33.3	36.0	38.5	40.8	42.9	45.0	46.9	48.7	52.0	54.9	57.5
2400	25.7	29.2	32.4	35.3	38.1	40.6	43.0	45.2	47.2	49.2	51.0	54.2	57.1	59.7
2500	27.7	31.3	34.7	37.7	40.5	43.1	45.5	47.7	49.8	51.7	53.5	56.8	59.6	62.2
2600	30.1	33.9	37.3	40.5	43.4	46.0	48.4	50.6	52.7	54.6	56.4	59.6	62.4	64.9
2700	33.0	37.0	40.6	43.8	46.7	49.4	51.8	54.0	56.1	58.0	59.7	62.8	65.5	67.9
2800	36.7	40.8	44.5	47.8	50.8	53.4	55.8	58.0	60.0	61.9	63.5	66.5	69.1	71.3
2900	41.4	45.7	49.5	52.8	55.7	58.4	60.7	62.8	64.7	66.4	68.0	70.8	73.2	75.2
3000	47.9	52.3	56.0	59.3	62.1	64.6	66.7	68.7	70.4	72.0	73.5	75.9	78.0	79.8
3100	57.3	61.5	65.0	68.0	70.5	72.7	74.6	76.2	77.7	79.0	80.2	82.2	83.8	85.2
3200	72.2	75.6	78.3	80.4	82.2	83.7	85.0	86.1	87.1	87.9	88.7	89.9	90.9	91.8

(t_2, Furnace gas exit temperature, F — leftmost column values)

Illustration #55 – Fuel Savings with Preheated Air. Courtesy of Combustion Handbook by North American Mfg. Co.
Note: An enlarged version of illustration 55 is shown in the Appendix on p. 215.

what savings might be made with preheated air, you first have to select a combustion air temperature. In our example, let's select 800°F (427°C) and a kiln maximum firing temperature of 2500° (1371°C). With these two conditions, the chart would indicate a savings of 34.7%, however, that is only the savings when firing at 2500°F. To obtain the real savings, one would have to integrate savings at each temperature with the specific firing cycle being considered. Since preheated air is a fairly costly addition to a kiln, a rule of thumb for batch kilns is that below a kiln temperature of 1350°C it usually doesn't pay, between 1350 and 1650°C, it can be justified by fuel savings. Above 1650°C it is necessary to have preheated air in order to make process temperature in any reasonable amount of time and also the fuel savings is very large. The most common means of producing the preheated combustion air in a batch fired kiln is with a stack recuperator system. Illustration 56 shows a typical stack recuperator system for a downdraft, batch fired kiln. In this illustration, the exhaust from the kiln comes from an underground flue and is pulled through ductwork to an exhaust fan. The recuperator is located in the ductwork connecting the underground flue to the exhaust fan. The recuperator is usually made of stainless steel and is a counter-flow shell and tube design. The combustion air fan blows cold air into the recuperator and this air leaves the recuperator preheated to some temperature. This is a very efficient system since the heat used to preheat is waste heat that has already left the kiln and was on its way to be exhausted. Stack recuperator systems should be equipped with a temperature controlled dilution air inlet upstream from the recuperator to prevent the recuperator from being overheated. Stack recuperator systems are set to operate at a maximum preheated air temperature controlled by the dilution air upstream from the recuperator.

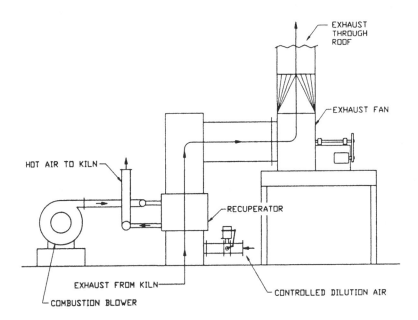

Illustration #56 – Stack Recuperator System. Courtesy of Swindell Dressler.

D. *Oxygen Enrichment*

Oxygen enrichment has the same effect on available heat and flame temperature as preheated combustion air. Specifically, replacing a percentage of the combustion air with oxygen increases the available heat and raises the flame temperature; however, oxygen is quite costly, therefore, it is not usually economic to use oxygen to save fuel. On the other hand, it is economical to use oxygen enrichment in small size kilns in order to reach higher temperatures such as 1800°C. Illustration 57 shows the effect of oxygen enrichment on available heat and flame temperature.

Available heat with various degrees of oxygen enrichment, and with standard air. This data is applicable only if there is no unburned fuel in the products of combustion. The average hot mix temperature may be read where the appropriate curve meets the zero available heat line. This chart is computer-calculated and corrected for dissociation for #2 fuel oil of Table 2.1. See footnote (†) on page 71.

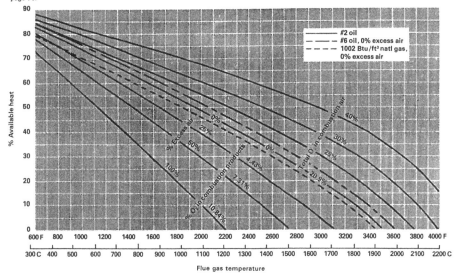

Illustration #57 Effect of Oxygen Enrichment on Available Heat and Flame Temperature. Courtesy of Combustion Handbook by North American Mfg. Co.
Note: An enlarged version of illustration 57 is shown in the Appendix on p. 216.

11. FUEL FIRED BATCH KILNS

11.1 Exhaust Systems — Batch Kilns

There are two ways to remove the exhaust from a fuel fired batch kiln; through the roof which is an updraft exhaust, or through the bottom of the kiln which is a downdraft exhaust. Illustration 58 shows a typical updraft exhaust system. With this system, most of the heating gases

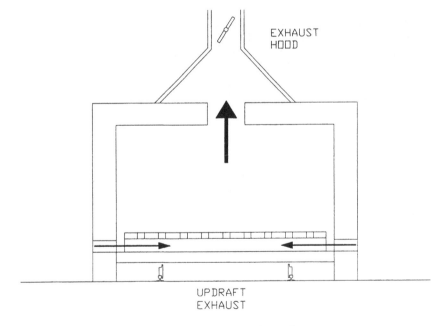

Illustration #58 – Updraft Exhaust System.

enter the kiln at the bottom and rise up through the load by bouyancy, exiting the kiln through a hole or flue in the roof. The major advantage of the updraft design is that it is less expensive to build than a downdraft design since it does not require any underground flues or an exhaust fan. However, the updraft design is only effective with light loads, tubes fired in a vertical mode, or hollow sanitaryware where the heating gases can easily rise up through the load.

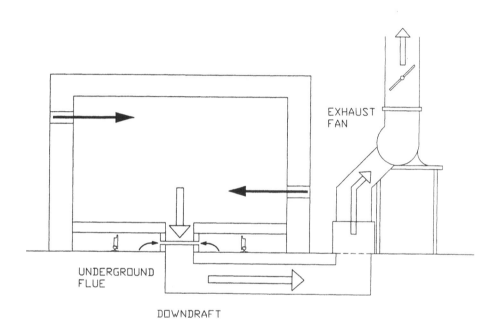

DOWNDRAFT

Illustration #59 – Downdraft Exhaust System.

Illustration 59 shows a cross section of a typical downdraft exhaust system. In downdraft batch kilns the burners are usually side mounted so the heating gases have to work their way around the perimeter of the load and then are pulled through the load by the force of the exhaust fan to the bottom of the kiln which is normally the coldest part of the kiln. The exhaust gases then travel through a refractory lined underground flue and then through sheet metal duct to an exhaust fan.

11.2 Fuel Fired Heating Systems — Batch Kilns

A. *Radiant Burner System*

Radiant burner systems such as flat flame burners are sometimes used in pottery kilns when gas heating is preferred to electric heating. Radiant burner systems like electric heating systems, require relatively small cross sections in the kiln because of the radiant heat transfer.

B. *Atmosphere Burner System*

Low velocity atmosphere burner systems are sometimes used in small pottery kilns, in cases where gas is preferred over electric heating. This type of system employs burners not unlike Bunsen burners located below the kiln so as to fire up through holes in the hearth. In this system, the low velocity heating gases rise up through the kiln by bouyancy and exit through the top of the kiln. A major advantage of this type of system is that it is very inexpensive because it does not require any type of a fan. The combustion air is inspirated by the gas pressure by Venturi action.

C. *Jet Firing with Controlled Excess Air*

The development of jet firing with controlled excess air in the late 1950's proved to be a major development which revolutionized the batch firing of ceramics and the way batch fired kilns would be built in the future. The overwhelming majority of fuel fired batch ceramic kilns today are fired with some type of jet firing system. The principle of jet firing is that high velocity gases coming from a high velocity burner into a kiln cause recirculation of the gases already in the kiln by entrainment. This recirculation in turn greatly improves the temperature uniformity throughout the kiln chamber.

Illustration 60 shows the basic jet burner laws. These laws are as follows:

1. Regardless of jet gas velocity, gas spread in a free medium from the jet source is always at a constant angle of 22°.
2. Decrease in gas velocity along the jet stream's axis is inversely proportional to the distance from the jet source.

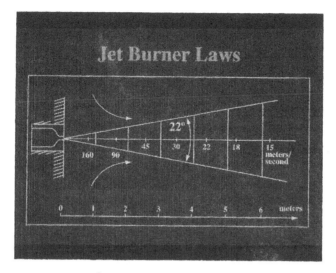

Illustration #60 – Jet Burner Laws.

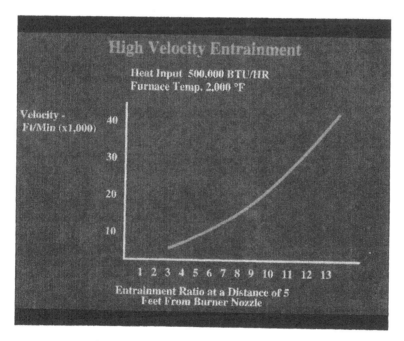

Illustration #61 – High Velocity Entrainment.

3. Temperature change along the jet stream axis is dependent upon the difference between injected gas temperature and the ambient kiln temperature. The decrease in temperature is also proportional to the distance from the jet source.

The fanning out of the burner jet, as shown in Illustration 60, is due to the inspiration and entrainment of ambient kiln gas into the jet stream. Illustration 61 shows how high velocity entrainment functions. As the burner jet velocity increases, entrainment of the ambient kiln atmosphere increases. As can be seen from this chart, at a nozzle velocity of 30,000 Ft/Min. (9100 M/Min.), the entrainment ratio is approximately 11:1. This means that for every cubic meter that enters the kiln through the burners, 11 cubic meters of ambient kiln atmosphere will be set into motion circulating within the kiln.

Illustration 62 shows the high velocity circulation in a kiln equipped with jet burners. As can be seen in this elevation view, burner jets cause secondary circulation within the kiln which greatly

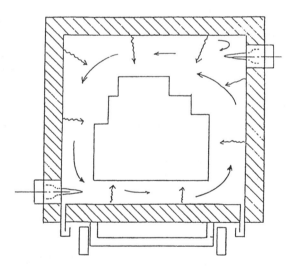

Illustration #62 – High Velocity Circulation, Elevation View.

improves temperature uniformity. Batch fired kilns, below 12 cubic meters in setting capacity, using the jet firing system, are very commonly perimeter fired. Illustration 63 shows the high velocity circulation pattern in a downdraft, perimeter fired kiln. As can be seen in the illustration, the jet burners fire in firelanes, which is a space around the perimeter of the load between the load and the walls of the kiln. The high velocity circulation around the perimeter of the load creates a very uniform temperature on the perimeter of the load, after which the pull of the downdraft exhaust system pulls the heating gases through the load and out the bottom of the kiln.

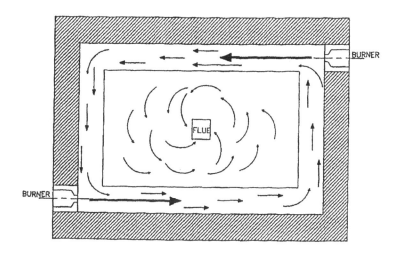

Illustration #63 – High Velocity Circulation Downdraft – Perimeter Fired.

Illustration 64 shows the high velocity circulation in a larger size jet fired batch kiln with a downdraft exhaust system. As can be seen in the diagram, what we have here is actually a series of perimeter fired kilns, all within one refractory envelope. Shuttle kilns up to 300 cubic meters in capacity can be built using this design. It should be noted that each setting area has its own exhaust flue.

The highest level of circulation and temperature uniformity is obtained when jet burners are firing at high levels of input between

50 and 100% of capacity; however, for the earlier parts of a firing cycle, if the burners fire between 50 and 100% of capacity, they will put in too much heat. Alternately, if in order to reduce the heat input into the kiln the burners are turned down, to their minimum level, the velocity is lost and the recirculation inside the kiln ceases. To solve this problem, jet burners are designed to fire with varying amounts of excess air up to 3000%. Illustration 65 shows the effect of excess air on burner jet temperature. At the beginning of the firing cycle when the kiln is cold, the jet burners operate with between 2500% and 3000% excess air which creates a resultant jet temperature of approximately 90°C (200°F). The burner jet temperature is then programmed up to 1800–1900°C by reducing the excess air level to a minimum. Most firing cycles call for between 1 and 2% free oxygen during the soak period which translates to approximately 10% excess air.

The first burner system to incorporate all of these features being jet burner velocities, coupled with independently controllable excess air, was patented in the U.S. in 1960. Illustration 66 shows this

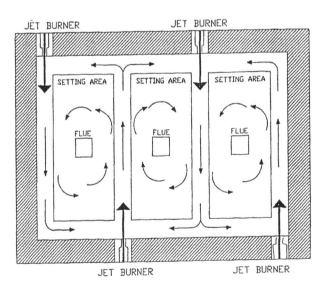

Illustration #64 – High Velocity Circulation Downdraft – Cross Fired.

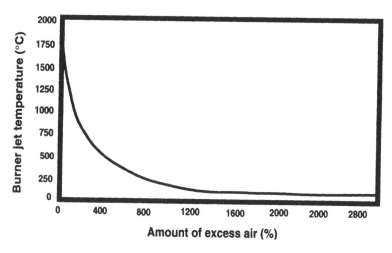

Illustration #65 – Effect of Excess Air on Burner Jet Temperature.

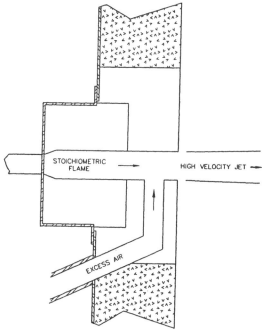

Illustration #66 – High Velocity Excess Air Burner with Excess Air Added After Combustion is Complete.

type of burner where the unique feature is that the excess air is added after combustion is complete. This type of burner was originally developed as a premix burner in order to obtain a short stoichiometric flame, more recent versions are now built with a nozzle mixing burner to improve the turndown. This type of jet burner with independently controllable excess air, is characterized by a hot burner block plus two fans and two air manifolds on the kiln. In other words, with this design there is a separate combustion air fan and combustion air distribution manifold plus a separate excess air fan and distribution manifold on the kiln.

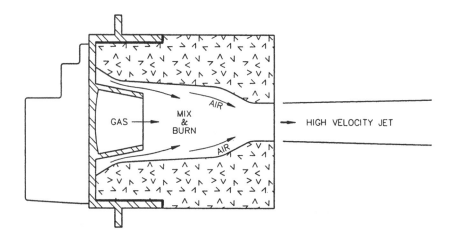

Illustration #67 – High Velocity Excess Air Burner with Staged Mixing.

In the early 1970's a new type of jet burner was developed that also had the ability to operate under very high levels of excess air. However, this burner used a different principle. The unique characteristic of this burner is the staged mixing of the gas and the air. Illustration 67 shows this type of burner. The air enters the burner block or combustion chamber, on the perimeter of the combustion chamber cavity. The gas enters on the centerline of the burner block cavity. The gas mixes with the combustion air at the center of the burner block and starts to burn. When firing at maximum fire, approximately 30% of the gas will be combusted inside the burner block and approximately 70% outside the burner block in the jet stream. Very high burner velocities are obtainable with this design due to the necked down relatively small opening in the burner block tile. This type burner can also turn down an incredible 100:1 on the gas without going out. Specific characteristics of this design are that the burner block remains relatively cold because it is constantly being swept with unburned combustion air. This cold burner block design is an advantage today in that it produces less nitrous oxides (NOx) than burners with hot burner blocks. These type burners can function with up to 3000% excess air and, therefore, produce burner jet temperatures as low as 90°C. These type burners are also characterized by the fact that they have only one fan for both combustion air and excess air and only one air distribution manifold.

D. *Pulse Firing – Jet Firing without Excess Air*

The newest innovation in jet firing is pulse firing which makes it possible to control the temperature in low temperature ranges without the use of excess air. With pulse firing, the burners operate at only two positions: 1) maximum fire — pulse or 2) minimum fire — low or off. The pulse time or period of maximum fire is one second or more. The heat input is controlled by varying the pulse time and the time between pulses. Illustration 68 shows an example of pulsing on a 48 second cycle where the heating capacity is varied by varying the on/off time. In this example, 6% of total heating capacity is achieved with a 3 second pulse and a 45 second gap between pulses. The pulse time is then increased to 6 seconds with a 42 second off time to obtain 13% of capacity. At 25% of capacity, there would be two 6 second pulses each separated by 18 seconds of off time so that the total off time would be 36 seconds. At 50% of capacity, there would be four 6 second pulses separated by four 6 second pauses. At 88% capacity there would be seven 6 second pulses separated by six 1 second pauses.

When burners are pulsing at maximum fire, the maximum burner velocity is achieved and maximum entrainment and circulation occurs.

Therefore, the major feature of pulse firing is that high velocity recirculation inside the kiln can be achieved without the use of excess air. By eliminating excess air, two things are achieved. First, there is a fuel saving since the excess air does not have to be heated and, secondly, by pulsing with a specific amount of excess air, the oxygen level can be precisely controlled at low temperature. Illustration 69 shows how the oxygen level can be controlled precisely when pulse firing. For example, when a ceramic body contains organic material for the purpose of burning that material out to control the porosity. If the organic burns too fast, the exothermic reaction can crack the ceramic. With excess air firing, it is difficult to control the oxygen much below 15% in the temperature range where organics burn out because if the excess air is reduced in order to reduce the oxygen, the heat input will be too great. With pulsing, for example, if the most advantageous oxygen level to control the burnout would be 7%, you simply would pulse with 50% excess air as can be seen on the chart in Illustration 69.

If you are firing graphite materials where zero oxygen is desirable, you would simply pulse with no excess air.

When pulse firing, if all the burners were pulsed at once and, subsequently went to their off or low position at once, there would be

Example: 48 second cycle

On Time	Off Time	Percent of Capacity
3	45	6%
6	42	13%
12	36	25%
18	30	38%
24	24	50%
30	18	63%
36	12	75%
42	6	88%
48	0	100%

Illustration #68 – Pulse Firing Capacity vs On/Off Time.

The oxygen level in the kiln can be controlled by pulsing with a specific amount of excess air.

Percent Excess Air in Pulse	Percent Free Oxygen in Kiln
0%	0.0%
10%	1.9%
50%	7.3%
100%	10.8%
500%	17.6%
1000%	19.1%

Illustration #69 – Oxygen Control with Pulsing.

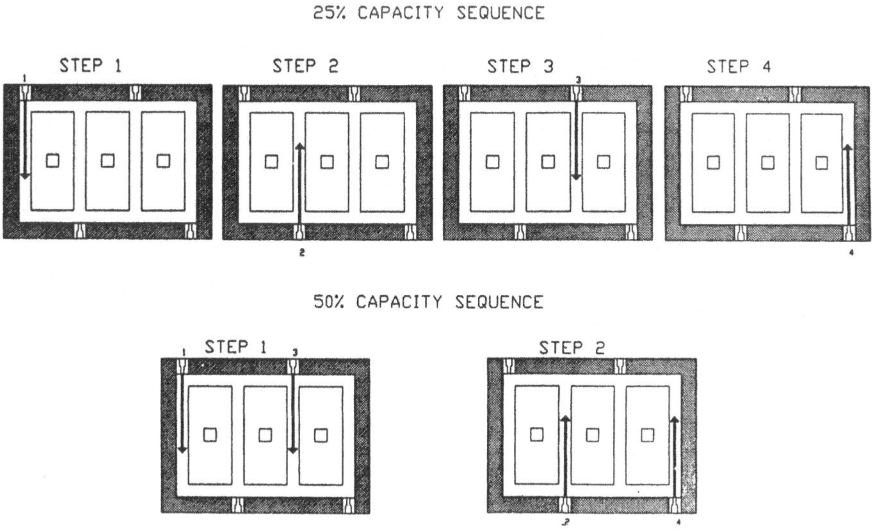

25% CAPACITY SEQUENCE

50% CAPACITY SEQUENCE

Illustration #70 – Burners Pulsed in Sequence to Minimize Pressure Fluctuations.

a substantial pressure swing in the kiln. To minimize the pressure fluctuation in the kiln, the burners are pulsed in sequence as shown in Illustration 70. This illustration shows a zone of four burners in a kiln which are all controlled from a single thermocouple. The example for 25% capacity shows that burner #1 first pulses, then burner #2, then burner #3, and then burner #4. With only one burner pulsing at a time, the maximum capacity under that condition is 25%. The 50% of capacity example shows burner #1 and #3 pulsing together, and then burner #2 and #4. In this example, the maximum capacity that could be obtained is 50%. At 100% capacity all four burners would be firing together. When a pulse system reaches 100% capacity, it is the same as a proportional system where the burners fire all the time.

E. *Oscillating Flame Front with Jet Burners*

The jet firing systems previously discussed in Section C and pulse firing in Section D are used with downdraft kilns where the burners fire into firelanes. In the case of updraft kilns, the burners fire underneath the load rather than into firelanes. In an updraft kiln, 80–100% of the burner capacity is located in the bottom of the kiln otherwise the bottom would be cold. Until the late 1970s, updraft batch kilns had to have a small cross section or else the temperature uniformity from side to side was very poor. The reason for this was that when the kiln was firing at low levels in the beginning, the underfired burners would tend to heat the near side more than the far side and the reverse was true when the burners were firing hard. In the late 1970s this problem was solved with the oscillating flame front system which permitted large cross section updraft shuttle kilns to be built with excellent temperature uniformity.

Oscillating flame front systems utilize the type of jet burner described in Illustration 67 which can be operated with a constant level of air (maximum) and the temperature is controlled by varying the gas only. Illustration 71 shows how the oscillating flame front system works. The total combustion air flow to the kiln is constant, however, it is distributed between the right and left sides of the kiln. Illustration A shows the condition where the combustion air is at

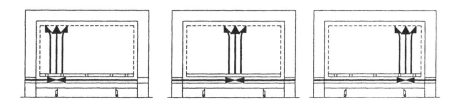

OSCILLATING FLAME FRONT

Illustration #71 – Oscillating Flame Front. Courtesy of Swindell Dressler.

minimum on the left side of the kiln. Illustration C shows the condition where the combustion air is maximum on the left hand side of the kiln and minimum on the right hand side. By alternately turning up the burners on one side of the kiln while turning the other side down, the point where the gases rise sweeps back and forth across the width of the kiln. With wide cross section kilns, the burners are directly opposed firing on the same centerline to maximize the upward pressure due to impingement. Smaller updraft kilns with a narrower cross section can be fired with unopposed burners sweeping back and forth.

11.3 Combustion Systems for Jet Firing

A. *Premix vs Nozzle Mix*

When jet burners were first developed in the late 1950s and early 1960s, premix burners were used. A premix burner is a burner where the gas and air is mixed to the correct ratio by a mixing machine or a mixer after which the gas mixture (10 parts of air and 1 part of natural gas) is delivered to all of the burners via a piping manifold. The advantage of the premix system is that the fuel air ratio being all mixed at one point is exactly equal at every burner; however, the disadvantages are that the burner can backfire and the turndown ratio is very poor (5:1). Backfiring is when the gas/air mixture burns back through the burner and into the piping system which obviously can cause serious damage. To prevent this, premix systems use fire checks and safety blowouts which are expensive equipment designed to arrest the flame in the pipe and shut the gas off. The poor turndown ratio of only 5:1 means that a burner rated at 250,000 kcal can only be turned down to 50,000 kcal.

Nozzle mixing burners were subsequently developed using cross connected ratio regulators. The nozzle mixing burners had the advantage that they could not backfire since the gas and air were only mixed together in the burner itself. As a result, the stoichiometric turndown was much better being approximately 10:1 which means that a 250,000 kcal burner could be turned down to 25,000 kcal. Later, with the development of the excess air type jet burners, as described in Illustration 67, the gas could be turned down by a ratio of 100:1 which means that a 250,000 kcal burner could be turned down to 2500 kcal.

B. *Proportional Control*

Proportional control is where the gas tracks the air in a proportional manner. For example, if the stoichiometric gas air ratio is 10:1 a proportional control system will maintain that ratio at every firing level. The mechanism for controlling the gas/air ratio in a proportional system is a gas ratio regulator in the gas line preferably at each burner which is cross connected with an impulse line to the air line going to that burner. When the zone air control valve opens or closes to change the air input to the burners in the zone the cross connected gas ratio regulators will automatically position the gas so as to maintain the gas ratio of 10:1.

Illustration 72 shows a typical combustion system for high velocity excess air burners. On this schematic diagram, valve (A) is the zone temperature control valve. This valve controls the combustion air to all burners in that particular zone. Gas regulator (C) is the gas regulator for an individual burner within the zone. Gas regulator (C) is connected to the combustion air line by an impuse line which is shown as a dotted line on the schematic diagram. There is no flow through the impulse line only pressure is transmitted through the impulse line. If the zone combustion air valve opens up calling for more heat, the increased air pressure will be transmitted to the diaphragm in the gas ratio regulator which will open up the regulator to allow more gas to flow in a proportional manner. In this proportional control scheme, motorized bleed valve (B) controls the excess air level. If no excess air in required, valve (B) is closed allowing the full impulse signal to go from the air line to the gas regulator. However, if excess air is required, valve (B) opens up bleeding off part of the signal which reduces the gas flow and in turn increases the excess air level. When valve (B) is wide open, the excess air level can be as high as 3000%.

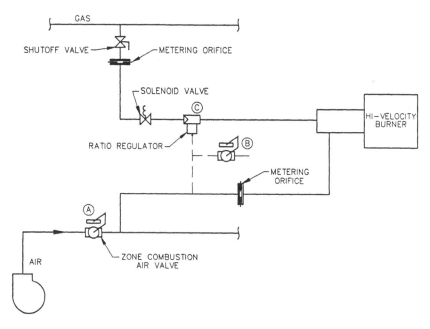

Illustration #72 – Combustion System Proportional Control or Gas Only Control. Courtesy of Hauck Mfg. Company.

C. *Gas Only Control*

With gas only control, the combustion air runs at a constant level and only the gas is varied to control temperature. Illustration 72 can also be used to illustrate gas only control. With gas only control, combustion air valve (A) does not exist and bleed off valve (B) becomes the zone temperature control valve. With this system, the combustion air is constant at maximum flow to all burners. In the beginning of the firing cycle when high excess air levels are desirable, the zone temperature control valve (B) is opened so as to bleed off most of the impulse signal and, therefore, put a very small amount of gas into the burner together with a high amount of excess air. As more and more temperature is required, control valve (B) gradually closes until stoichiometric firing is achieved.

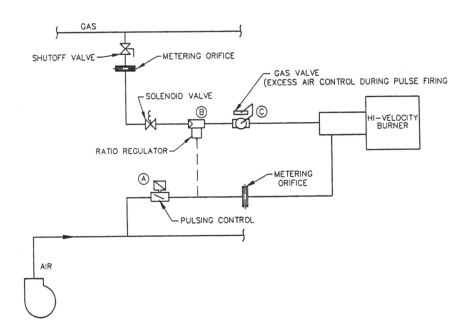

Illustration #73 – Combustion System Pulse Firing. Courtesy of Swindell Dressler.

D. *Pulse Control*

Combustion systems for pulse control are quite a bit different since instead of turning the burners up and down, they are turned on and off, or high/low. Illustration 73 shows the combustion system for pulse firing. With this system, there is a combustion air pulse valve for each burner. In this case, it is labeled (A) and there is a gas/ratio regulator for each burner which is labeled (B). The gas regulator is cross connected with an impulse line to the air line so that when the air valve opens up to maximum flow the correct amount of gas for maximum air will also flow to the burner. The zoning of burners in a pulse system is done electrically by linking all the pulse air valves for one zone together. One of the advantages of pulsing is that the burner zoning can be changed simply by changing wires. There are various ways to control the excess air level in a pulsing system; however, in the schematic shown in Illustration 73, the excess air is controlled by valve (C) which reduces the gas flow below the proportionate level if excess air is required.

E. *Pulse Proportional Control*
In 1992 the pulse proportional control system was introduced which is the most versatile firing system for a batch kiln developed to date and is the current state-of-the-art. In this system, the high velocity excess air burner can be fired on a proportional basis with all burners on at all times and controlled excess air or alternately it can be pulsed to provide control of the oxygen level at any given kiln temperature.

F. *Oscillating Flame Front*
The combustion system for an oscillating flame front kiln using high velocity excess air burners is shown in Illustration 74. With this system there is a constant air flow to the kiln at all times from the air fan, however, the air is distributed from one side of the kiln to the other by the oscillation valves (A). As the air level rises and falls on each side of the kiln, the gas tracks the air through ratio control regulators (B) one for each burner. The temperature control valve for each zone of burners is bleed-off valve (C), which provides fuel only throttling.

11.4 Types of Fuel Fired Batch Kilns

A. *Box Kiln*
A box kiln is a kiln with a solid bottom. In other words, it does not have a car bottom. Box kilns have limited use in the ceramic industry,

however, they can be used in situations where cars are not feasible and the down time to load and unload the kiln is not a hardship. Box kilns can be updraft or downdraft and cost less to build than a car bottom kiln. Box kilns are often used in situations where the product can be loaded in and out of the kilns with a fork truck. Illustration 75 shows a typical fuel fired box kiln.

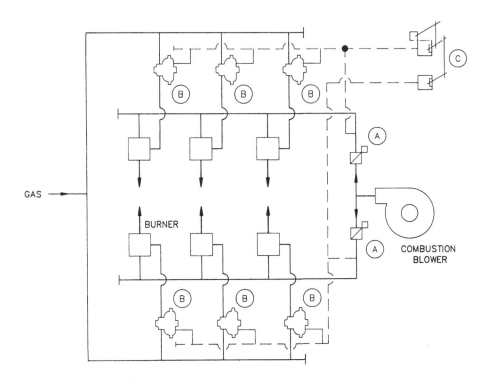

Illustration #74 – Combustion System – Oscillating Flame Front. Courtesy of Swindell Dressler.

B. *Elevator Kilns*

An elevator kiln is a kiln that is permanently mounted up in the air on steel support legs so that the bottom of the kiln, which is a car, can be lowered down out of the kiln and then rolled on a track out from under the kiln for loading and unloading. Elevator kilns can be supplied with two kiln cars so that as soon as the fired car is removed an unfired car can be put back into the kiln to start the next firing. Elevator kilns are pretty much relegated to small sizes such as one cubic meter setting volume or less where they are slightly cheaper to build than a bell kiln. In larger sizes, bell kilns are less expensive to build than elevator kilns and have the advantage that the bell kiln is at ground level in the firing position whereas with an elevator kiln, one must climb up a ladder to look into the kiln or build a platform around the kiln at the elevated level. Elevator kilns can be built updraft or downdraft and they are most commonly used as pilot plant kilns or kilns to fire small ceramic parts.

Illustration #75 – Box Kiln. Courtesy of Swindell Dressler.

C. *Bell Kilns*

A bell kiln is a kiln usually mounted on two hydraulic rams so the entire kiln can be elevated above the top of the car load. The bottom of the kiln is a car which runs on a track the same as with an elevator kiln. Bell kilns can be built in quite large sizes with setting volumes of up to 50 cubic meters. The reason for this is that with lightweight refractories the kiln bell actually weighs less than the car fully loaded in many cases. Illustration 76 shows a fuel fired bell kiln in the load/unload position with the bell elevated above the load. Bell kilns are always downdraft with the major advantage over shuttle kilns being that they have no doors. The elimination of the door is especially important in high temperature kilns (1500°C and higher) where door sealing and door maintenance becomes a problem. The bell design has the additional advantage in that it gives excellent support to the lightweight refractory walls with a wrap around steel shell on all four side walls. The bell design also makes it easier to transport the kiln with the refractory lining installed. This means that it is easier to factory build the kilns and ship them around the world and it also improves the resale value of the kilns.

Illustration #76 – Bell Kiln in Load/Unload Position. Courtesy of Swindell Dressler.

Illustration 77 shows the bell kilns in firing position with the kilns at ground level. Bell kilns are almost always supplied with multiple cars so that there is no down time for loading and unloading. The kiln loading and unloading time is generally around 30 minutes and follows the following sequence:

Illustration #77 – Bell Kiln in Firing Position. Courtesy of Swindell Dressler.

1. Kiln bell is raised on hydraulic rams above the fired kiln load.
2. The fired kiln car is moved out from under the kiln bell, usually with a transfer car, and shifted to a parallel track.
3. A car carrying green ware is then moved via the transfer car and placed underneath the bell.
4. The kiln bell is then lowered by gravity and when in position is restarted for the next firing cycle.

D. *Shuttle Kilns*

A shuttle kiln is a batch fired kiln with one or two doors and a car bottom. Small shuttle kilns will have one kiln car as the entire bottom of the kiln whereas large shuttle kilns will have a whole train of kiln cars. Shuttle kilns can be built 6 to 7 meters wide, up to 6 meters high, 20 meters long or larger; there are no practical limits. Shuttle kilns with a usable setting height of 5 to 6 meters have been built to fire 6 meter high electrical porcelain insulators for use in power plants. Shuttle kilns used to refire sanitaryware are quite often 6 to 7 meters wide.

Illustration #78 – Two Shuttle Kilns. Courtesy of Swindell Dressler.

Illustration 78 shows two shuttle kilns with the doors down in firing position. These shuttle kilns happen to have vertical lift doors, however, doors can be side sliding or pivot, if headroom is a problem. Single door shuttle kilns, as shown in Illustration 78, utilize a transfer car to shift the kiln cars, whereas double door shuttle kilns can be built with a straight piece of track running through with the shuttle kiln located in the middle of the track with two sets of cars. One set of cars is always on one side of the kiln or in the kiln and the other set of cars is on the other side of the kiln or in the kiln. With this system a transfer car is not required.

Illustration #79 – Downdraft Shuttle Kiln. Courtesy of Swindell Dressler.

There are two types of shuttle kilns: updraft and downdraft. Illustration 79 shows a downdraft shuttle kiln with the door open and the car removed to expose the underfloor flues. Downdraft shuttle kilns have firelanes as can be seen in this photograph, the burners are on and indicate that there are four firelanes in the kiln. This kiln also has three kiln cars, each with its own flue.

Illustration 80 shows an updraft shuttle kiln with the door open so you can see one of the updraft flues in the roof. This kiln has an oscillating flame front firing system which underfires the load with opposed burners sweeping back and forth. The shelving which holds the sanitaryware, in this case, has slits in it to allow the heating gases to rise up through the shelves and through the load to the roof of the kiln. As can be seen in the photograph, this kiln has an all fiber lining.

Illustration #80 – Updraft Shuttle Kiln. Courtesy of Swindell Dressler.

12. FUEL FIRED CONTINUOUS KILNS

12.1 Fuel Efficiency in a Continuous Kiln

In general, modern continuous kilns use about half the fuel of modern batch kilns firing the same product on the same cycle. The two most important reasons continuous kilns use less fuel are as follows:

1. The walls of a continuous kiln do not have to be heated every cycle since they remain at temperature sometimes for years. Therefore, all the heat stored in the walls of a continuous kiln is saved when compared to a batch kiln where heat must be replaced every cycle.
2. The most significant reason for the increased efficiency in a continuous kiln is due to the fact that the ware moves through a continuous kiln counter-flow to the exhaust gases. As the exhaust gases pass over the incoming ware, they transfer their heat so that the amount of heat lost up the stack is much less in a continuous kiln than in a batch kiln.

Illustration 81 shows the heat transfer in a continuous kiln. As can be seen, the flow of exhaust gases from the hot zone moving from right to left is towards the PC fan or products of combustion exhaust fan. The car travel is from left to right. With this system, the exhaust gases are pulled around and through the incoming product as they travel down the tunnel towards the entrance end, thereby transferring their heat to the incoming product. PC fan temperatures as low as 120°C are common. PC fan temperatures below 100°C are not desirable since that can cause rain in the entrance end of the kiln. By comparison the exhaust temperature in a batch kiln tracks the firing cycle temperature so that when the kiln is at

113

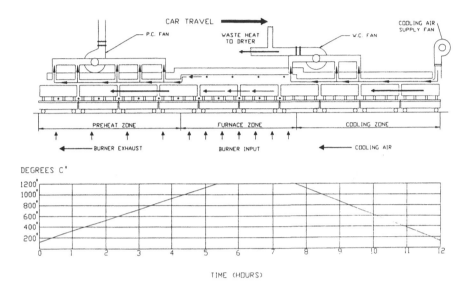

Illustration #81 – Heat Transfer in a Continuous Kiln.
Note: An enlarged version of illustration 81 is shown in the Appendix on p. 217.

1000°C the exhaust temperature is approximately 1000°C. When the kiln is at 1300°C, the exhaust temperature is approximately 1300°C.

Illustration 81 also shows the flow pattern in the cooling zone which is also counter-flow to the car travel. In this diagram, the cooling air which mainly enters from the exit end of the tunnel, is pulled counter-flow over the fired ware by the WC fan or ware cool exhaust fan. The exhaust from the WC fan is clean, hot air which may be used to supply the heat requirements in a dryer or, in some cases, to help heat the building. Illustration 81 shows one of three possible situations for the cooling zone of a continuous kiln; that being, where all of the cooling air is pulled out of the tunnel by the WC fan. The other two situations which will be discussed later in this chapter are where part of the cooling air is allowed to continue into the hot zone as preheated combustion air and alternately where all of the cooling air is allowed to proceed into the hot zone to provide all of the combustion air.

The time/temperature curve in Illustration 81 shows the temperature of the ware as it moves through the continuous kiln with respect to time. This time/temperature curve is the firing cycle for the product being fired. Continuous kilns are usually designed for a specific firing curve, however, the top temperature and throughput rate can be varied.

Illustration 82 shows the kiln pressure in a continuous kiln. In this diagram, the horizontal line running through the center of the kiln is the zero pressure point so that below the horizontal line it is a suction. As can be seen, the maximum pressure is at the exit end of the kiln at the cooling air supply fan. The pressure is reduced at the ware cool fan as much of the cooling air is removed from the tunnel. The burner input in the firing zone tends to hold the pressure positive; however, the strong pull of the PC fan pulls the kiln pressure into the negative range in the preheat zone. Kiln pressure continues to get more and more negative as it approaches the front end and the PC fan intakes. Since the flow is always from high pressure to low pressure, this diagram shows the flow through the tunnel in a typical continuous kiln.

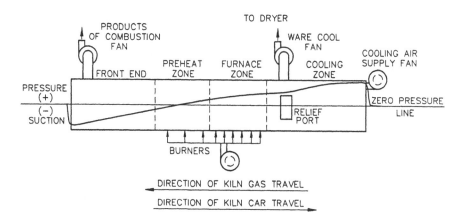

Illustration #82 – Kiln Pressure in a Continuous Kiln.

12.2 Tunnel Kilns

A. *Car Conveyance*

The most common type of continuous kiln is the tunnel kiln which uses a train of refractory lined kiln cars to carry the ware through the kiln. The kiln cars run on a track through a refractory lined tunnel. The train of kiln cars is slowly moved through the tunnel by a hydraulic pusher ram located at the entrance end. Illustration 83 shows a typical tunnel kiln from the entrance end. The hydraulic power unit for the pusher ram can be seen adjacent to the kiln cars which are entering the kiln tunnel. The return track is parallel and to the right of the kiln in this photograph. The cars loaded with green ware are moving toward the transfer car. The transfer car is shown in a pit and is used to transfer the kiln cars from the return track to the kiln track. Also, this photograph shows the PC fan located on the floor adjacent to the kiln on the right hand side.

Illustration #83 – Tunnel Kiln with Kiln Cars. Courtesy of Swindell Dressler.

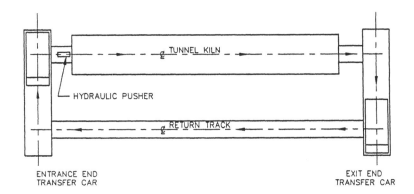

Illustration #84 – Floor Plan Layout, Typical Tunnel Kiln.

Illustration 84 shows the floor plan layout for a typical tunnel kiln arrangement. As can be seen in this diagram, the kiln car movement through the tunnel kiln is from left to right where the fired kiln cars are then moved via the exit end transfer car to the return track. The car movement on the return track is from right to left. With this arrangement, the fired ware would be offloaded from the kiln cars on the return track and greenware would be reset onto those empty kiln cars for return to the kiln. The cars loaded with greenware would be transferred to the kiln track via the entrance end transfer car. The hydraulic pusher at the entrance end of the tunnel kiln would then push the cars loaded with unfired ware into the entrance end of the tunnel. Since there is a continuous car train inside the tunnel kiln, every time a car with unfired ware is pushed into the entrance end, a car with fired ware is pushed out of the exit end.

Modern tunnel kilns often have a completely automatic car moving system controlled by PLCs. When a fired kiln car exits the tunnel kiln, it trips a limit switch and an automatic transfer car at the exit end moves over and pulls the kiln car onto itself with a hydraulic ram built into the transfer car. The transfer car then shifts to the return track and discharges the fired kiln car onto the return track using the same hydraulic pusher ram. The kiln cars are moved along the return track with a cable haulage system built into the floor. The automatic transfer car at the entrance end of the kiln will be set into motion when a space is available for a kiln car at the entrance end of the kiln. This vacancy is again sensed by a limit switch after which the entrance end transfer car moves to the return track and pulls a kiln car loaded with greenware onto itself. The transfer car then switches to the kiln car track and discharges the kiln car onto the kiln track at the entrance end to fill the vacancy.

B. *Sizes of Tunnel Kilns*

Tunnel kilns come in a wide variety of sizes. Therefore, for discussion purposes in this book, kilns will be categorized to size as follows:

(1) Small size tunnel kiln — less than 30 meters long

(2) Mediun size tunnel kiln — 30 to 60 meters long

(3) Large size tunnel kiln — more than 60 meters long

C. *Continuous Push Tunnel Kilns*

(1) *Firing Pattern Over and Under*

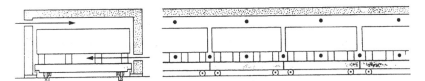

Illustration #85 – Continuous Push Tunnel Kiln Firing Pattern.

There are two types of tunnel kilns; the first being a continuous push tunnel kiln where the car train is pushed continuously at a uniform rate of so many meters per hour. In a continuous push tunnel kiln the burners fire over and under the load as shown in Illustration 85. The bottom burners fire underneath the bottom shelf of product and the top burners over top of the load. This design has the advantage that it is easier to get heat into the bottom of the product load since the burners fire underneath the load itself. Due to the fact that heat rises, the great majority of the burners in this type kiln are located to fire underneath the load. The bottom burners start at the entrance end of the tunnel. Top burners are often located on a wider spacing between burners than bottom burners and are also often located only in the hot zone.

The disadvantage of this design is that the product load must be supported on a refractory superstructure to permit the burners to underfire the load. With light kiln loads, the superstructure itself can be light in weight where with heavy loads the superstructure naturally gets heavier. The heavier the superstructure, the more energy must be used just to heat the superstructure and the more the cost of the superstructure. Therefore, continuous push kilns are often used to fire lighter kiln loads and index push kilns, which have no superstructure, are used to fire heavier loads.

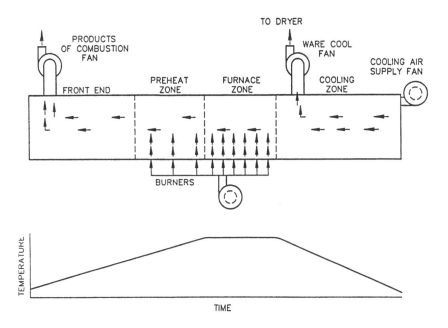

Illustration #86 – Longitudinal Flow, Continuous Push Tunnel Kiln.

Illustration 86 shows the longitudinal flow in a continuous push tunnel kiln. As can be seen in this diagram, the cooling air enters the exit end of the tunnel and exits through the WC fan with as little as possible flow from the cooling zone to the hot zone. This flow pattern offers the maximum flexibility of firing cycle in a tunnel kiln since the heating section of the kiln is separated from the cooling section with regard to tunnel flow.

(2) *Advantages of the Continuous Push System*

The major advantages of the continuous push system are as follows:

a. Excellent temperature uniformity from top to bottom due to the fact that the burners underfire the load.
b. Faster firing cycles are possible due to the fact that the heat up of the bottom part of the load is not held back by the mass of the kiln car.
c. Flexibility of firing cycle and firing cycle profile.

(3) *Firing Systems for Continuous Push Kilns*

a. High velocity premix systems — high velocity premix burner systems have been popular with some European kiln builders due to the fact that they are more fuel efficient than most other systems. The reason they are fuel efficient is because very little excess air is used and therefore, fuel does not have to be expended to heat up the excess air. However, this system has three disadvantages.

The first disadvantage is that since there is a combustible gas/air mixture in the piping, the flames can backfire into the piping system and therefore, backfire protection must be provided at every burner.

The second disadvantage in premix burner systems is that they are subject to relatively large pressure swings in the kiln since a 10% change in heat requirement causes more change in flow down the tunnel. With gas-only systems, which will be discussed later, a 10% change in heating requirement would require only a 1% change in flow down the tunnel.

The third disadvantage is that in order to maintain velocity if the heat requirement in a section of the tunnel is low, then small size burners with smaller size burner orifices must be placed there. If it is desired to change the firing cycle, it may be necessary to actually change the burner sizes. This is not the case with other systems.

b. Jet Firing — gas only control. Jet firing with gas only control when used on a tunnel kiln uses the same type of burners that are used in batch kilns with the same system. This type burner was described in Chapter 11.2, Illustration 67 and Chapter 11.3, Illustration 72 which shows the schematic for the combustion system. Since this system uses controlled amounts of excess air to develop circulation and temperature uniformity, it will use somewhat more fuel than firing systems that do not employ excess air. However, the advantages of this system are substantial. Since these burners are nozzle mix burners where the gas and air only come together inside the burner, they cannot backfire into the piping manifold. The air flow at each burner is set to be constant and only the gas is varied to control heat input. This means that the flow down the tunnel is very constant and pressure control becomes quite easy. With a 10% change in heat requirement there is only a 1% change in flow down the tunnel since there are 10 parts of air for every one part of gas flowing down the tunnel and the air does not change.

Lastly, this system gives excellent temperature uniformity due to the jet velocities generated at the burners even with low heat input due to the excess air exiting from the burner block along with the combustion gases. It is also very flexible as burner sizes do not have to be changed to maintain velocities.

c. Jet firing with oscillating flame front. This is the same system described in Section 11.2 e, Illustration 71. Oscillating flame front firing systems have been used in very wide underfired continuous fired kilns to improve side to side temperature uniformity. This system is also a gas only control system as described in Section b, above. All the advantages of jet firing with gas only control also apply here. The system is usually employed with tunnel kilns having a setting width of 3 meters or more.

d. Jet firing with pulse — this is the same system that was described in Section 11.2 d, as jet firing without excess air. In a tunnel kiln this system has the advantage that it can save fuel by eliminating the need for excess air to obtain high velocity circulation.

(4) *Applications for Continuous Push Tunnel Kilns*

Continuous push tunnel kilns are used for a wide variety of applications from spark plugs, and advanced ceramics, to whitewares and structural clay products. Some typical examples are as follows:

Illustration #87 – Continuous Push Dinnerware Tunnel Kiln. Courtesy of Swindell Dressler.

Illustration 87 shows a continuous push tunnel kiln firing dinnerware. This particular kiln is 65 meters long and fires 20,000 dinner plates per day at a temperature of 1250°C.

Illustration 88 shows a typical continuous push tunnel kiln firing sanitaryware. This kiln is 55 meters long and fires 1036 pieces of sanitaryware per day on a 14 hour cycle at 1230°C.

Illustration #88 – Continuous Push Sanitaryware Tunnel Kiln. Courtesy of Swindell Dressler.

Illustration 89 shows a continuous push tunnel kiln firing ceramic tile. This kiln is 60 meters long and fires 500,000 sq. meters of tile annually.

Illustration 90 shows a continuous push tunnel kiln firing facebrick. This kiln is 73 meters long and has a useful setting width of 4.4 meters. It fires facebrick in 19.8 hours with an output of 97,000 per day. Fuel consumption is 400 kcal/kg gross.

Illustration #89 – Continuous Push Ceramic Tile Tunnel Kiln. Courtesy of Swindell Dressler.

D. Index Push Tunnel Kilns

1. Firing Pattern — Fires in Firelanes

As opposed to continuous push tunnel kilns where the car train is continuously advanced thru the tunnel, in index push tunnel kilns the car train is indexed the distance between firelanes. In other words, if the firelanes are on 1.5 meter centers, the entire car train is advanced 1.5 meters each push. After which it will be stationary for a period of time. The length of dwell time between pushes is dependent upon the speed of firing and also the size of the kiln. A typical example could be a 100 meter long tunnel kiln firing on a 48 hour cycle with firelanes on 1.5 meter centers. In this case there would be 42 minutes between pushes and each push would last approximately one minute, during which time the entire car train would be advanced a distance of 1.5 meters.

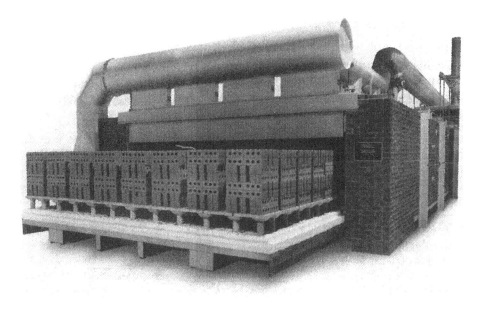

Illustration #90 – Continuous Push Facebrick Tunnel Kiln. Courtesy of Swindell Dressler.

Illustration #91 – Tunnel Kiln Car with Firelanes.

Illustration 91 shows a tunnel kiln car for an index pushed tunnel kiln with firelanes. There are two types of index pushed tunnel kilns depending on whether the burners fire into the firelane from the sides of the kiln or the top of the kiln. Illustration 92 shows an index push tunnel kiln with a side fired burner pattern. In this example, there are two Burners high within a given firelane and the burners within a given firelane fire from opposite sides of the kiln. With index push tunnel kilns that have low setting heights, usually there is only one burner in each firelane, with the burners firing from opposite sides on alternate firelanes.

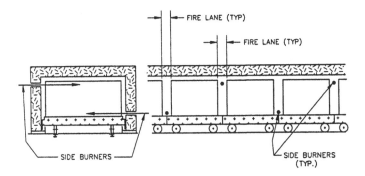

Illustration #92 – Index Push Tunnel Kiln, Side Fired Burner Pattern.

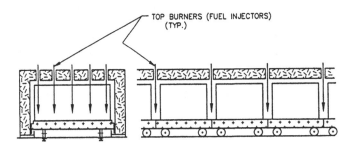

Illustration #93 – Index Push Tunnel Kiln, Top Fired Burner Pattern.

A top fired index push tunnel kiln is shown in Illustration 93. In this design the number of burners in a given firelane is dependent upon the width of the kiln. As can be seen in the diagram, the burners fire from the roof down to the car deck within the firelane. In most top fired tunnel kilns the top burners are really not burners at all, but instead are fuel injectors. In other words, most of the air for combustion is pulled up the tunnel and the roof mounted fuel injectors insert the fuel plus some air into the air stream in the firelanes where combustion takes place. Top fired tunnel kilns can operate on powdered coal, oil or gas.

D.2 *Advantage of Index Push*

The advantage of the index push design is that it can handle heavy loads and high temperatures since no support refractories are required to create a space underneath the load for burners to fire in. Index push tunnel kilns can be built for temperatures as high as 1850°C. Another advantage of the index push tunnel kiln is that when top fired with fuel injectors so that most of the combustion air is preheated coming from the cooling zone, maximum fuel efficiencies can be achieved.

D.3 *Firing Systems for Side Fired Index Push Tunnel Kilns*

a. *Jet Firing with Gas Only Control*

This is the same system as described under continuous push tunnel kilns, Section C.3, sub. b. The longitudinal flow down the tunnel with this system is the same as previously described in Illustration 86. This system has the advantage that it is very flexible with regard to firing cycle since there is very little flow between the cooling zone and hot zone. The disadvantage being that fuel consumption is sacrificed since preheated combustion air is not used.

b. *Preheated Air in a Side Fired Index Push Tunnel Kiln*

Illustration 94 shows the longitudinal flow pattern in an index push side fired tunnel kiln. Part of the combustion air is coming direct from the cooling zone as hot preheated air. With this system, fuel efficiency is increased over using no preheated air but firing cycle flexibility is compromised in that long soak periods are not possible. For example, when firing basic refractory brick, in some cases a long soak period can be advantageous. To achieve this end and still use a high percentage of preheated combustion air to get efficiency, the kiln can be built as shown in Illustration 95. In this design there is very little flow between the cooling zone and the hot zone inside the tunnel itself. However, clean, hot air is taken from the cooling zone and supplied to the burners as combustion air either through a hollow wall plenum or an arch plenum. This design is more costly to build but offers more flexibility with the firing cycle.

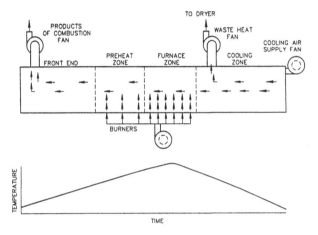

Illustration #94 – Preheated Combustion Air Direct from Cooling Zone.

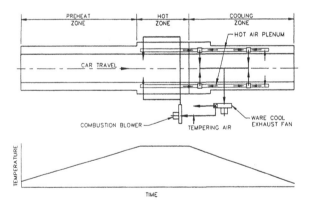

Illustration #95 - Preheated Combustion Air Indirect from Cooling Zone.

D.4 *Firing Systems For Top Fired Index Push Tunnel Kilns*

Illustration 96 shows the longitudinal flow pattern in a top fired tunnel kiln with 100% of the combustion air coming from the cooling zone as preheated air. This type of tunnel kiln will use the lowest amount of fuel per ton of product fired. However, there is very little flexibility of firing cycle profile and there is no flat soak period at the top.

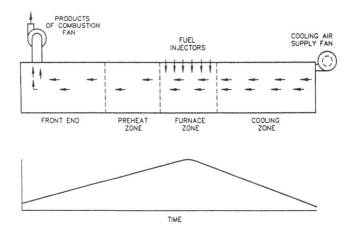

Illustration #96 – 100% of Combustion Air from Cooling Zone.

D.5 *High Pressure Designs*

Normal tunnel kilns operate with an overall pressure drop from the exit end to the entrance end of approximately 5 mm of water column (0.2 of an inch water column). High pressure designs have a pressure drop of ten times as much or 50 mm water column (2 inches water column pressure). The high pressure designs can be index push kilns fired from the side or fired from the top. However, in either case, the side walls and the roof have to be encased in gas tight steel in order to hold the high pressure in the kiln. Some designs have been developed to use a water seal between the car train and the kiln walls to hold in the high pressures. The whole purpose of the high pressure design is to reduce fuel consumption and force flow of heating gases through the load. However, these designs are very inflexible with regard to firing cycle and are very sensitive to change of setting. A rule of thumb is that the higher the pressure drop in the tunnel kiln, the more sensitive the kiln is with regard to maintaining the firing cycle profile. Increased fuel economy is obtained in the high pressure designs because the clearances between the interior walls of the kiln and the product are very tight and therefore the velocities down the tunnel are very high which forces the heating gases through the load and improves the heat transfer. Obviously it requires a very high pressure drop in order to pull the heating gases through the load with the tight clearances.

D.6 *Applications for Index Push Tunnel Kilns*

One major application for index push tunnel kilns is in the refractory field due to the high temperatures that are required there. As previously stated, the index push tunnel kiln has the advantage that it does not require a superstructure on top of the kiln cars to support the load so that the burners can underfire the load. The higher the firing temperature the more important this aspect of index push tunnel kilns is. The second major use of index push tunnel kilns is for firing structural clay products where due to the low cost of the product itself, the fuel consumption is much more important. Both the top fired designs that use 100% of the combustion air from the cooling zone of the tunnel kiln and the high pressure designs are extremely fuel economic.

Illustration 97 shows a high temperature index push tunnel kiln firing basic refractories. This kiln is side fired and uses some preheated combustion air to obtain elevated flame temperatures and increase efficiency. Illustration 98 shows a top fired index push tunnel kiln firing face brick.

Illustration #97 – Side Fired Index Push Tunnel Kiln. Firing Refractories – Courtesy of Swindell Dressler.

Illustration #98 – Top Fired Index Push Tunnel Kiln Firing Face Brick – Courtesy of Hans Lingl Gmbh.

12.3 Roller Hearth Kilns

A roller hearth kiln is a kiln that has a conveyor to move the product through the kiln instead of kiln cars. The roller hearth is a continuous kiln the same as a tunnel kiln and the air flow pattern inside the roller hearth kiln is the same as a tunnel kiln in that most of the heat is put into the center of the kiln and the exhaust gases are pulled toward the entrance end, counterflow to the product movement. The cooling zone of a roller hearth works the same as in a tunnel kiln in that cooling air is pulled from the exit end toward the center of the kiln where it is normally pulled out with a ware cool fan. The conveyor itself is made of ceramic rolls or stainless steel rolls depending on the temperature. The most popular type of ceramic rolls are mullite and recrystalized silicon carbide.

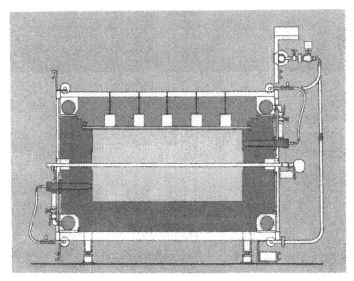

Illustration #99 – Cross Section Roller Hearth Kiln. Courtesy of SITI S.p.A.

Illustration 99 shows the cross section through a typical roller hearth kiln. Burners are usually arranged to fire both underneath and above the roll conveyor. The major advantage of a roller hearth kiln over a tunnel kiln is that since there are no kiln cars which must be heated and cooled and essentially only the product or the product and a single piece of kiln furniture move through the kiln, extremely fast firing cycles are possible. Generally speaking, roller hearth firing cycles range from as fast as 30 minutes up to three or four hours. When firing cycles get longer than four hours it is usually less expensive to use a one high tunnel kiln design with light weight cars.

Roller hearth kilns are expensive to build per meter of length because of all the conveyor mechanism and the cost of the ceramic rolls normally used in the hotter areas of the kiln. Roller hearth kilns are also limited in width which is dictated by the length of the roll which has to pass through both side walls. Generally speaking, a setting width of 2.0 meters is close to a maximum. With setting height being limited to one high, a setting width being limited to approximately 2.0 meters, the only way to get capacity is to fire very fast.

There are several types of roll drive mechanisms with a chain drive being the least expensive to build and the gear drive being the most precise. Illustration 100 shows a close up view of a gear drive. This design has the advantage that the ceramic rolls are free to expand and contract and are easy to replace without interrupting the overall operation of the conveyor. The gear drives where the ceramic roll is fixed to a metal end which in turn is connected to gears, is the most popular type of drive today because it is more precise and does not slip.

Illustration #100 – Gear Type Roll Drive. Courtesy of SITI S.p.A.

Illustration 101 shows a fully automatic roller hearth kiln.

Illustration #101 – Roller Hearth Kiln. Courtesy of SITI S.p.A.

Illustration 102 shows a close up view of the setter tiles holding the small alumina ceramic parts. The setter tiles are conveyed through the roller hearth kiln by the driven roll conveyor, and the parts to be fired are placed on top of the setter tiles.

The most successful use of roller hearth kilns has been in the ceramic tile industry where most size tiles can be fired directly on the rolls without the requirement for any kiln furniture. This is extremely important since it eliminates the cost of the kiln furniture and more than halves the weight that has to be heated in order to fire the product.

Illustration 103 shows a typical roller hearth kiln for firing tile. Today, the majority of wall tile in the world is fired in roller hearth kilns.

Illustration #102 – Roller Hearth Firing Small Alumina Parts on Setter Tiles – Courtesy of Swindell Dressler.

Illustration #103 – Roller Hearth Kiln Firing Tile. Courtesy of Swindell Dressler.

12.4 Rotary Kilns

Rotary kilns are used for calcining ceramic oxides which is the process of heating mineral materials to a temperature below their melting point for the purpose of driving off volatile materials, oxidizing or reducing. A rotary kiln is fundamentally a large refractory lined pipe which is on a slight angle from horizontal and slowly rotates. The raw material is fed into the elevated exhaust end of the pipe so that the material literally walks through the pipe counterflow to the exhaust gases. The hot raw material drops out of the exit end of the rotating pipe into a cooling section which is usually located at a level below the main rotary kiln. Rotary kilns usually have one large burner located at the exit end as shown in Illustration 104.

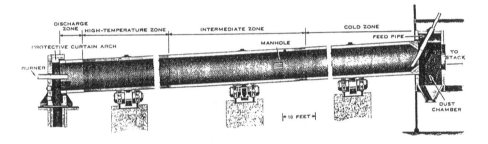

Illustration #104 – Cross Section of Rotary Kiln. Courtesy of Modern Refractrory Practice by Harbison Walker.

13. HOW TO SIZE A KILN

13.1 Basic Criteria

Before someone can size a kiln, we first must have the basic criteria about the kiln, which are as follows:

A. The product to be fired – abrasives, refractories, sanitaryware, face brick, etc.

B. The firing temperature – reference Part I, Illustration 17, entitled "Typical Firing Temperature of Ceramics".

C. Type of fuel to be used – reference Part II, Section 7.1, Fuel versus Electric.

D. Batch or continuous – reference Part II, Section 7.2, Batch versus Continuous.

E. Quantity of product required per year. This item can be given in weight such as kilograms or number of pieces such as 30,000,000 bricks per year.

F. Setting density in kilograms per cubic meter or pounds per cubic foot. Or in the case of face brick, the setting density could be expressed as a specific setting pattern from a setting machine. Setting density is the actual amount of product in a specific volume of kiln space. For example, grinding wheels which can be made out of silicon carbide have a density of 2500 kilograms per cubic meter. However, because the grinding wheels are fired on shelving and there must be air spaces for the gases to pass in and around the wheels, the total weight including the kiln furniture, shelving and the wheels that can be placed in a cubic meter in a kiln is approximately 1000 kilograms. However, out of the 1000 kilograms, only 600 kilograms are actually abrasive wheels or the product being fired, the balance being kiln furniture and sand. Therefore, the load density in this case would be 600 kilograms per cubic meter. This load density would be used to determine the cubic meters of kiln space required to fire a specific quantity of abrasives per year.

TYPICAL PRODUCT LOAD DENSITIES

PRODUCT	ACTUAL PRODUCT LOAD DENSITY
Abrasives	600 kg/cubic meter
Fire Clay Refractory Brick	1200 kg/cubic meter
Basic Refractories	1600 kg/cubic meter
Sewer Pipe	300 kg/cubic meter
Electric Insulators	250 kg/cubic meter
Sanitaryware	130 kg/cubic meter

G. *Firing Cycle Time*

What we mean by this is the total time in to out of the kiln, often referred to as the cycle cold to cold.

13.2 Example Calculations of Kiln Size

Using the basic criteria outlined in Section 13.1, we will calculate the setting volume and the actual usable setting dimensions on the car for a batch fired kiln in Example 1. In Example 2 we will calculate a tunnel kiln size for firing Face Brick, however, in this case instead of having a load density in kilograms per cubic meter we will use a setting pattern determined by an automatic brick setting machine.

Example #1

Basic Criteria

      a. Abrasives

      b. 1300°C

      c. Natural Gas

d. Batch
e. 750,000 kg/year
f. 600 kg/cubic meter
g. 72 hours (3 days)

750,000 kg ÷ 360 days	= 2083 kg/day
2083 kg/day × 3 day cycle	= 6250 kg/load
6250 kg ÷ 600 kg/M³	= 10.42 M³ Setting Space Required

Ideal Size for a One Car Bell Kiln

Determine Car Size
1. Choose Setting Height   1.6 M (Convenient Height for Loading)

2.   <u>Setting Volume</u>   <u>Setting Height</u>   <u>Setting Area</u>
     10.42 M³   ÷   1.6   =   6.5 M²

3. <u>Choose Setting Width</u> – 2 Meters (Heating gases must travel only
                                  1M to get to center of Load)

4.   <u>Setting Area</u>   <u>Setting Width</u>   <u>Setting Length</u>
     6.5 M²   ÷   2.0   =   3.25 M

Setting Dimensions for Car
     Setting Width   = 2.0 M
     Setting Length = 3.25 M
     Setting Height = 1.6 M

Example #2   Continuous Kiln

Basic Criteria
a. Face Brick (204 × 102 × 64 mm)
b. 1100°C
c. Natural Gas
d. Continuous
e. 36,000,000 per year

f. Setting Pattern – 80 pcs/Bung

                    Bung Dimensions W – 408 mm
                                            L – 408 mm
                                            H – 816 mm

g. 24 hours

    36,000,000 pcs ÷ 360 days   = 100,000 pcs/day

    100,000 × 1 day Cycle   = 100,000 pcs in kiln

This capacity dictates a large tunnel kiln. In this case we will choose an over and under fired continuous push design.

Determine Car Size

1. Car Setting Height – One Bung High = 816 mm
    (Lower setting height gives better
    temperature uniformity top to bottom)

2. Car Setting Width   8 Bungs Wide × 408 mm = 3264

                                      7 Spaces × 50 mm   = <u> 350</u>
                                                       3614

    (Jet burners permit kilns this wide which reduces kiln length and saves cost)

3. Car Length   6 Bungs Long × 408 mm = 2448

                                      6 spaces × 75mm   = <u> 450</u>
                                                        2898

    (Car length is Arbitrary)

    Car Size

        Setting Width   3600 mm
        Setting Length   2900 mm
        Setting Height   816 mm

Determine Kiln Length

1. No. pieces/car – 48 Bungs × 80 pcs/bung = 3840 pcs per car

2. 100,000 pcs in Kiln ÷ 3840 pcs/car = 26 cars in Kiln

3. 26 cars x 2.9 M/car = 75.4 M Kiln Overall Length

14. KILN CONTROLS

14.1 On/Off Control

On/off control is the simplest form of control and is still used today for some small electric kilns with metalic elements. Illustration 105 shows how on/off control operates. This type system usually operates with a dead band such that the system does not shut off until the sample temperature exceeds the set point by a few degrees and does not turn back on again until the sample temperature drops below the set point by a few degrees. The spread between the turn off point and the turn on point is the dead band and this creates a sign wave type oscillation above and below the set point.

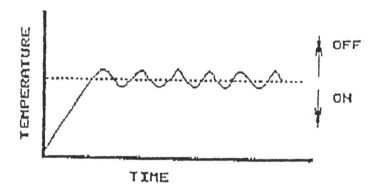

Illustration #105 – On-Off Control.

14.2 P.I.D. – Proportional Integral Derivative Control Logic

The overwhelming majority of all kilns today use P.I.D. control for temperature control. P.I.D. is an improvement over on/off control which eliminates the dead band oscillation by applying a corrective action proportional to the deviation from the set point. Illustration 106 shows how P.I.D. control works to track the set point temperature without oscillation. This diagram shows what happens in a kiln if the actual temperature falls below the set point temperature. With P.I.D. control the heat input would be increased proportionate to the deviation from the set point by the actual temperature. As the actual temperature approaches the set point temperature, the P.I.D. system anticipates that there will be an overshoot and starts to cut back the heat input in advance of reaching the set point. By this method, the overshooting is eliminated and the process temperature comes in exactly to the set point temperature and tracks thereafter.

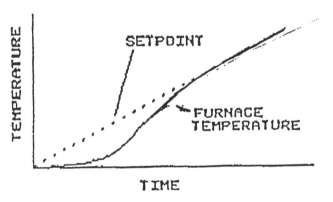

Illustration #106 - P.I.D Control.

14.3 Actuators

A temperature control system has to have a means of increasing or decreasing the heat input to the kiln. For electric kilns the means for controlling the heat input to the kiln are quite different than for gas fired kilns.

A. *Electric Kilns*

(1) *Contactors*

Contactors which are basically switches can be either mechanical or mercury type and they are turned on and off by the temperature control system to vary the heat input to the elements. Time proportioning can be used with contactors, however, only metallic heating elements can be used with contactors. Contactors are usually only used on small kilns with metallic elements.

(2) *Silicon Control Rectifiers (SCR Units)*

SCR units are solid state and therefore last a lot longer than mechanical switches and also operate differently in that they use voltage proportioning for control. Illustration 107 shows a typical SCR power control unit, the kind that would typically be used to control an electric kiln. The recommended type of SCR unit to use with either silicon carbide or molybenum disilicide elements is a phase angle or synchronous SCR power control unit with transformer in-rush protection and a current limit feeding transformer. The SCR-Transformer combination with a multi-tap transformer compensates for element aging. The phase angle SCR provides proportional control on the transformer primary. For element resistance changes with temperatures, such as with molybdenum disilicide elements, an SCR unit with a current limit feeding transformer is required. Since the molybdenum disilicide element needs approximately three seconds to heat up, this is called a soft start. The SCR power control unit operates on the transformer primary to provide the soft start.

B. *Fuel Fired Kilns*

(1) *Motorized Valves*

The actuators used on fuel fired kilns are motorized combustion air or fuel valves, which are controlled by the temperature control system. If heat input is required to an individual burner or group of burners the temperature control system will cause the motorized valve which controls that burner or group of burners to open, thereby putting more heat into the kiln. The motor operators used on control valves are proportional motors with a slide wire to provide a closed loop feedback, to the control instrument.

On pulse fired systems the heat input is controlled by pulsing a solenoid combustion air valve open and closed. The frequency of open time determines heat input.

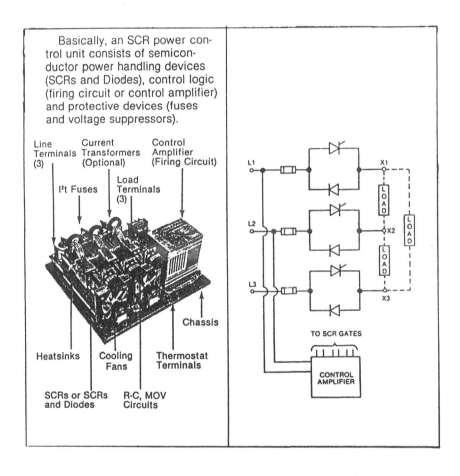

Illustration #107 – SCR Power Control Unit.

14.4 Sensors

For each item that is to be controlled in a kiln there must be a sensor. Different types of sensors used in a kiln are as follows:

A. *Temperature Sensors*

The most common type of temperature sensor is a thermocouple, with the three most common type of thermocouples being chromel-alumel, which is Type K, and is normally used up to 1200°C. 10% platimun rhodium, Type S, used to around 1450°C, and 6–30% platinum-rhodium Type B, used to 1760°C. For temperatures above 1760°C rayotubes are used which focuses the temperature in the kiln thru a lens onto a thermopile.

B. *Pressure Sensors*

The most widely used type of pressure sensor today is a transducer which measures the pressure in a kiln and converts it to a millivolt signal. The microprocessor type or computer type control instrumentation, that controls the temperature will also control the pressure thru a separate loop.

C. *Fuel Consumption Sensors*

Fuel consumption is normally measured by totalizing gas meters which measure gas flow by metering orifices or vortex meters. The data is converted to a millivolt signal which can be sent to the instrumentation system for recording.

D. *Oxygen Sensors*

The sensors used to measure oxygen level in the kiln atmosphere are oxygen analyzers. The most common type uses a zirconia tube with a platinum catalyst. Oxygen analyzers can convert their signal to a millivolt signal and send this signal to the control instrumentation for recording or control.

E. *CO and CO_2 Sensors*

The most common sensor to measure reducing atmospheres in kilns is a $CO + CO_2$ analyzer which measures the $CO + CO_2$ level in the kiln and converts it to a millivolt signal which can be sent to the control instrumentation for recording or control.

14.5 Basic Microprocessor Controllers

Microprocessor controllers are electronic controllers which combine a digital programmer and a digital P.I.D. controller in a compact unit, shown in Illustration 108. This is the most common type of controller used on batch fired kilns where the set point must be programmed up and down to create the firing cycle. The particular unit shown in Illustration 108 will store up to 39 different programs, and also has 12 event outputs to trigger alarms or energize auxiliary control devices. Single loop microprocessor controllers which are P.I.D. controllers that operate to a single set point are often used on continuous kilns, with one controller for each zone of burners.

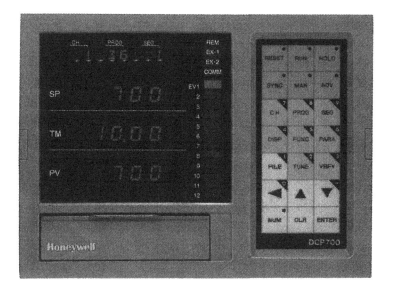

Illustration #108 – Microprocessor Program Controller. Courtesy of Honeywell.

14.6 PLC Controllers

Programmable Logic Controllers can also be used to control kiln temperatures, pressures, etc. However, PLCs are more commonly used in the Ceramics Industry to control car moving systems which can require a very high number of separate events to be controlled. PLC control systems usually cost more than the computer control systems discussed in Section 14.7.

14.7 Computer Control Systems

In the late 1980s a new type of furnace control system was introduced that married computer data acquisition capabilities together with loop and logic control microprocessor controller systems. These new computer control systems store firing data from previous firings for instant retrival. Illustration 109 shows a computer control system for a modern Batch

Illustration #109 – Computer Controls for A Batch Kiln. Courtesy of Swindell Dressler.

Fired Kiln. Programming is done from a keyboard at the PC located on a table adjacent to the control panel. Firing data is recorded on the printer located adjacent to the PC on the table. Manual control microprocessor controllers are mounted on the control panel as backup to the computer system. The manual controllers are also used for initial furnace adjustment and trouble shooting. This control system controls the following systems:

(1) Program control of temperature in multiple zones.

(2) Program control of excess air (Oxygen Level)

(3) Automatic kiln pressure control to a set point

(4) Automatic flame supervision on all burners

(5) Automatic program control of cooling with the burners off

(6) Automatic shut down at the end of the cycle

(7) An annunciator system in the case of kiln malfunction with the malfunction being identified.

(8) Fuel usage

(9) Record all firing data

Illustration #110 – Large Computer Control Center. Courtesy of Swindell Dressler.

Illustration 110 shows a large computer control center for a tunnel kiln, dryer and car moving system.

Computer systems for Tunnel Kiln Plants can keep track of individual cars as they move through the Plant, Dryer and Kilns. The thermal history of an individual car can be tracked and recorded.

PART III
FIRING PRACTICE BY INDUSTRY

Part III discusses the current state of the art in firing practice for each industry. The chapters in this section are arranged by industry and in alphabetical order.

15. ABRASIVES INDUSTRY

15.1 Abrasives Industry

The Abrasive Industry produces two types of grinding wheels, namely resin bonded grinding wheels and vitrified grinding wheels. Since resin bonded grinding wheels are not fired, this chapter will deal only with the vitrified type, which represent approximately 50% of the total volume manufactured. Vitrified grinding wheels vary in size from small mounted points for jewelers, up to large pulp stones which can be up to two meters in diameter. The great majority of vitrified grinding wheels are made out of silicon carbide or fused alumina grains as the abrasive material.

The reason that ceramic bonds are used in grinding wheels is that the ceramic bond is a high temperature bond and can withstand the significant temperatures that are developed during the grinding process. The ceramic bond itself is the key item which controls the performance of the grinding wheel since it controls the wear rate of the wheel. A well designed grinding wheel is a self-sharpening tool as the ceramic bond holds the cutting grain just long enough until its sharp cutting edge has worn down, and then allows that grain to be released, thereby exposing a new sharp grain to continue the cutting. If the ceramic bond is too weak, the grinding wheel will wear away too quickly. On the other hand, if the ceramic bond is too strong, the grinding wheel will become dull and cease to cut.

Most abrasive companies will have a variety of different bonds to handle different grinding situations. Some of the bonds are engineered to be soft, some are engineered to be hard. The firing temperature for the various bonds range from approximately 1100°C up to 1300°C. However, the biggest factor that affects the firing cycle is the size of the wheels being fired, which means that very small wheels could be fired in hours, whereas very large wheels could require firing cycles as long as eight to ten days.

15.2 Firing Problems

The major firing problems encountered when firing abrasives are as follows:

Preheat — During the preheat period carbonaceous materials will volatilize out of the product as CO, CO_2 and H_2O. The carbonaceous materials come from temporary binders and pore formers which are meant to burn out. Sufficient time in the cycle must be allowed for the carbonaceous materials to evolve from the center of the wheel and escape from the surface of the wheel prior to the time that any bond would develop and therefore, block passage of gases. If the surface of the wheel closes down before all the carbonaceous materials have been removed, a black core will be developed inside the wheel. This black core will be made of carbon and will make the wheel a reject.

Another problem that can occur during the preheating period is that if the temperature uniformity is poor, heating cracks can develop due to differential thermal expansion between the inside of the wheel and the surface of the wheel. Therefore, the heating rate for a large wheel must be significantly slower than for a small wheel to permit time for the heat to be conducted from the surface of the wheel to the interior of the wheel.

Soak — The soak period permits the bond to be fully developed throughout the entire thickness of the wheel. It is extremely important that uniform density is developed since grinding wheels turn at very high rpms and would tend to fly apart if they had dense spots within their perimeter. Therefore, temperature uniformity during the soak period is the most important single criteria for a kiln which fires grinding wheels.

Cooling — The early part of the cooling, from the end of the soak period down to the annealing range is relatively non-critical and can be done fairly rapidly. However, when cooling thru the period from 750°C down to 450°C, where the bond hardens, the temperature uniformity throughout the wheel must be quite good to avoid the possibility of cooling cracks. It is the general practice when firing larger grinding wheels, to fire down through this period. In other words, with the burners firing, the cooling is accomplished by mixing large quantities of excess air with the combustion products to provide a more gentle cooling than would be possible without the burners being on. Below 450°C, the cooling rate can often be accelerated and the burners can be turned off.

Typical Firing Cycles for Abrasives

Illustration 111 shows three typical firing cycles for abrasives. The fast cycle is a 24 hour cold to cold firing cycle for small wheels. The intermediate cycle, which is 72 hours cold to cold, is a typical cycle for firing intermediate size wheels. The long cycle shown here has a total firing time of 144 hours and would be used to fire large wheels. As can be seen from these firing curves. the lengths of the firing cycles is primarily dependent on the size of the wheel being fired.

15.3 The Most Popular Type of Kiln for Firing Abrasives is the Bell Kiln

Prior to the development of the Bell kiln in the late 1950's and early 60's, most grinding wheel plants would have one or two large tunnel kilns to fire all of their wheels. The problem with this was that the tunnel kilns had to operate on a firing cycle based on the largest wheel being fired in the tunnel kiln. The large wheels usually only represented about 5% of the total volume in the plant. In other words, 95% of the wheels were fired on a cycle length longer than necessary.

When tunnel kilns were replaced with a bank of Bell kilns, it was not only possible to fire each size group of wheels on the optimum firing cycle but it was possible to fire them all simultaneously, using different Bell kilns. For example, in one given week an abrasive plant with four Bell kilns could fire four one day cycles for small wheels, four three day cycles for intermediate wheels, and two 6 day cycles for large wheels.

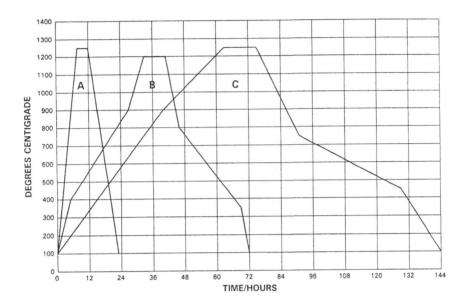

Illustration #111 – Typical Firing Cycles for Abrasives.

The Bell kilns were down draft design and used high velocity jet firing systems to create excellent circulation and temperature uniformity throughout the entire firing cycle. As a result of these additional features, kiln firing losses were greatly reduced as compared to the older tunnel kiln designs.

Today, better tunnel kiln designs are available that can provide greatly improved temperature uniformity over those of thirty years ago. However, the concept of multiple Bell kilns within a given abrasive plant is still the most efficient way to go because of the great variety of firing cycle requirements within each plant. Bell Kilns also provide flexibility with regard to market requirements. Market flexibility means that a given plant may have to produce a disproportionate amount of large wheels during a time period to fulfill a large order, and then switch directly from that to a disproportionate amount of small wheels to fulfill a large order in that category. Changes like this are very difficult with continuous tunnel kilns where they are quite easy to handle with a series of batch fired Bell kilns.

The third reason why Bell kilns are preferred over tunnel kilns for firing abrasives is that the abrasive product is quite expensive per pound and the fuel cost to fire it as a percentage of the cost of the product is very small, usually less than one percent. The major advantage of the tunnel kiln over a batch fired kiln is lower fuel consumption and the major disadvantage is lack of flexibility. Therefore, since in the abrasive industry flexbility is much more important than fuel consumption, the flexible batch firing system wins out.

Illustration #112 – Bell Kiln Firing Abrasives. Courtesy of Swindell Dressler.

16. ADVANCED CERAMICS

Advanced Ceramics is a catchall term that applies to ceramic parts that are made for their electrical, chemical, thermal or mechanical properties. From a firing standpoint, Advanced Ceramics can be divided into two general categories, oxides and non-oxides. The oxides are usually fired in gas fired kilns in an oxidizing atmosphere and the non-oxides are usually fired in electric kilns in an non-oxidizing atmosphere.

16.1 Some Typical Examples of Advanced Ceramics in the Oxide Category are as follows:

1. Alumina
2. Mullite
3. Steatite
4. Forsterite
5. Cordierite
6. Aluminum Titanate
7. Strontium Titanate
8. Barium Titanate
9. PZT
10. Zirconia
11. Magnesium Oxide
12. Beryllium Oxide

16.2 Some Typical Examples of Advanced Ceramics in the Non-Oxide Category are as Follows:

1. Silicon Carbide
2. Boron Carbide
3. Silicon Nitride
4. Aluminum Nitride
5. Titanium diboride

158

16.3 Firing Problems

Advanced Ceramics – Oxides

Most advanced ceramics products tend to be small in size and therefore can be fired in hours rather than days. Also due to the small product size and high cost per kg of product, the production runs are relatively small in volume. These factors therefore dictate that the kilns be relatively small in volume which favors Bell kilns and roller hearths over large volume shuttle and tunnel kilns.

The two major firing problems usually have to do with controlling burnout reactions and final temperature uniformity which controls physical size and properties.

Advanced Ceramics — Non-oxides

The major firing problem with most non-oxide advanced ceramics is that we live in a sea of oxygen and these products must be protected from oxygen during the firing process. They are also fired at very high temperatures in the 2000 to 2500°C range depending on the product. The special atmosphere dictates an electric kiln. However, the only types of electric elements that can handle these very high temperatures are Tungsten and Graphite. Since Graphite is much less expensive, it is most commonly used.

16.4 The Most Popular Types of Kilns for Firing Oxide Type Advanced Ceramics are Bell Kilns and Roller Hearth Kilns

The Bell kiln is very popular for firing oxide type advanced ceramics because it is so flexible. The Bell kiln can fire one product at 1100°C on one day and a totally different product to 1700°C on the following day. Bell kilns can fire cycles as fast as twelve hours cold to cold or very long cycles lasting many days if the product requires it. The most common sizes of Bell kilns that are used in the advanced ceramics industry are between one and ten cubic meters of usable setting space. In these size ranges, Bell kilns usualy cost less to build than tunnel kilns of similar capacity. Lastly, Bell kilns offer excellent temperature uniformity with their down draft exhaust systems and their multizone high velocity burner systems.

Illustration 113 – Bell Kiln with Recuperator for Firing Advanced Ceramics to 1760°C. Courtesy of Swindell Dressler.

If advanced ceramics product can be fired in four hours or less, and if there is sufficient volume to keep a continuous kiln running twenty four hours a day, the roller hearth is a very interesting choice. The roller hearth being a conveyor type of kiln, lends itself to automation and because the product is usually fired only one layer high, the roller hearth normally provides excellent temperature uniformity. Since the roller hearth is not nearly as flexible as the Bell kiln, it normally would be used in situations where only one product is being fired at one temperature.

Most oxide type advanced ceramics are not made in sufficient quantities to warrant a Tunnel kiln, but there are a few exceptions. Some exceptions are: catalyst carriers for the Chemical and Petroleum Industry and honeycomb substrates for automobile catalytic converters.

Illustration #114 – Roller Hearth Firing Small Alumina Parts on Setter Tiles. Courtesy of Swindell Dressler.

16.5 The Most Popular Type of Equipment for Firing Non-Oxide Ceramics is a Vacuum Furnace

The combination of the very high temperature requirement coupled with the atmosphere requirement makes the vacuum furnace ideally suited for non-oxide advanced ceramics. The most common size furnaces used have work spaces varying from 300 mm × 300 mm × 300 mm to 300 mm × 450 mm × 1000 mm.

These furnaces use the vacuum system to evacuate the air after which the protective atmosphere is introduced. These furnaces generally have a water cooled outer shell and use graphite elements and graphite felt type insulation. Some typical temperatures and atmospheres are as follows:

Material	Firing Temperature	Atmosphere
Silicon carbide	2300°C	Argon
Boron carbide	2500°C	Argon
Silicon nitride	2000°C	Nitrogen
Aluminum nitride	2000°C	Nitrogen
Titanium diboride	2300°C	Argon

Illustration #115 – Typical Vacuum Furnace for Firing Non-Oxide Advanced Ceramics. Courtesy of Thermal Technology.

17. CARBON PRODUCTS

17.1 High Temperature Baking

Carbon products (Graphite) include the electrodes that are used for electric melting of steel, the graphite anodes that are used in the process of making aluminum metal, as well as many other graphite products that are used as bearings, heating elements, etc. The largest volume part of the carbon industry are electrodes for the steel and aluminum industry which are mainly baked in ring pit furnaces at temperatures of 1000°C. However, recently a higher temperature baking process has been developed for non-electrode type graphite products. This high temperature baking process requires temperatures in the 1260 to 1320°C range and utilizes ceramic type kilns for the baking process. The baking process is where tar or pitch is converted to pure carbon by driving off the hydrogen and other volatile materials from the tar. This process is slow and requires rather long firing cycles as shown in Illustration 116. Illustration 116 shows a seven day bake cycle and a 23-1/2 day bake cycle.

The process starts by making a cylinder out of tar which is then packed along with additional cylinders in a stainless steel canister or a refractory crucible along with sand. The sand is used to support the cylinders so that they will maintain their shape during the baking process. The purpose of the stainless steel canister is to not only support the pieces together with the sand but also to maintain a very strong reducing atmosphere inside the canister, since the baking process must take place in complete absence of oxygen. The larger the individual pieces of carbon being baked and the larger that the canisters are which contain the pieces, the longer the baking cycle must be. Illustration 116 shows a seven day bake cycle that would be used for baking small parts fired in relatively small canisters such as 600 mm in diameter by 600 mm high. Firing cycle B which is a 23-1/2 day cycle would be used for larger carbon pieces fired in larger canisters which might be 1200 mm in diameter by 1500 mm high.

The recent development of pulse firing has shown itself to be ideal for the high temperature graphite baking process. Traditional low temperature baking processes below 1000°C use stainless steel recirculating fans inside of the furnaces to get temperature uniformity, whereas with the high temperature baking process, the temperatures are too high for internal fans. Therefore, circulation is achieved with pulse firing jets. The pulse firing system has the ability to create circulation without the use of excess air and therefore in the absence of oxygen.

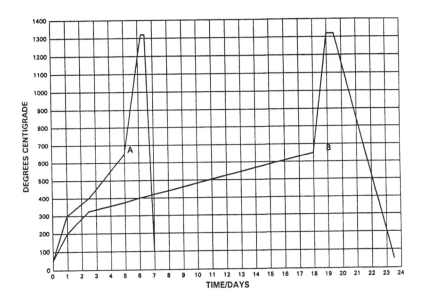

Illustration #116 – Typical Firing Cycles for High Temperature Graphite Baking.

17.2 The Most Popular Type of Kilns for the High Temperature Carbon Baking Process are Pulse Fired Shuttle Kilns

As can be seen from Illustration 116, the baking cycles are quite long and due to these long cycles, economics would dictate a rather large kiln. Since bake cycles vary in length of time dramatically, depending on the size of the pieces being baked, a batch kiln which can fire different bake cycles every time is much more practical than a continuous kiln. The most ecomonical large size batch kiln is usually a shuttle kiln, therefore, the trend in this industry today is for relatively large pulse firing, down draft shuttle kilns. The down draft exhaust is usually preferred because it helps in obtaining good temperature uniformity as well as provides an economic means to burn up the smoke which emanates during the process, using an afterburner built into the down draft flue system.

18. CLAY PIPE

The major use for clay pipe in the world is for underground sewer systems, whereas during the last 20 years the main competitor which is PVC Pipe, has taken a larger and larger share of the world wide market, due to the fact that the PVC pipe costs less to buy. Some towns and cities still do not permit PVC pipe due to the fact that PVC pipe can fail from deflection if not installed properly, and does not have as good resistance to chemical degredation as clay pipe.

18.1 Clay vs PVC

However, the clay pipe industry may be on the verge of a big comeback due to new technology for installing the pipe, namely micro tunneling. Micro tunneling eliminates the need for digging up the street by boring underneath the street using a boring mole. This installation technique requires that the pipe itself have very high compressive strength. Since clay pipe has a compressive strength much higher than PVC, 95% of all pipe used in micro tunneling in sizes 600 mm and down, is clay. This process has been used extensively in Europe and Japan and is now starting to be used in the United States as well. With micro tunneling the total installed cost of the sewer line underneath the road with clay pipe is less than the cost of PVC pipe installed in the conventional manner by digging up the street.

18.2 Firing Problems

Clay pipe are typically made around the world in sizes from 100 mm diameter to 1200 mm diameter and in lengths up to 2 meters long. In the United States clay pipe are made from 4" diameter to 42" diameter and in 5 and 6 foot lengths. Clay pipe is actually quite a difficult product to make because it has to meet technical requirements for performance strength, porosity, etc. However, economics are such that a pre-engineered raw material cannot be used. In other words, the clay pipe manufacturer must work with the red clay that is available and still hold a uniform set of technical specifications for the finished product. A typical firing cycle in a tunnel kiln, firing pipe up to 600 mm in diameter, by 2 meters high, is 60 hours cold to cold. Firing cycles in shuttle kilns vary from 42 hours to 90 hours cold to cold, depending on the size of pipe being fired.

18.3 The Most Popular Type of Kilns for Firing Clay Pipe are Tunnel Kilns and Shuttle Kilns

The most economical way to fire clay pipe is in a tunnel kiln, however, due to the high setting height required (2 meters), the tunnel kilns must be of necessity, very long, typically 150 meters or more. Therefore it takes a substantially large plant output to be able to keep a tunnel kiln running with a full load at all times. The type of tunnel kiln that is most suitable for pipe is an underfired, continuous push, oscillating flame front type of kiln where opposed burners create an upflow of heating gases through the pipes so the pipes are heated from both inside and outside simultaneously. With this type of kiln the point of burner impingement is oscillated back and forth across the width of the kiln to provide uniform heat flow up through the pipes, from side to side. As previously stated tunnel kilns usually fire pipes only up to 600 mm in diameter, with the larger diameter pipes being fired in shuttle kilns.

The most popular type of shuttle kiln for firing clay pipes is an updraft design with an oscillating flame front firing system the same as is used in the tunnel kiln. With this system the pipes are heated from inside and outside simultaneously as the heating gases flow up through the pipes.

In a tunnel kiln plant several shuttle kilns would also be used to fire the pipe diameters in excess of 600 mm. Whereas in smaller plants, where there is insufficient capacity to keep a large 150 meters long tunnel kiln operating, that plant would only have shuttle kilns. The "all" shuttle kiln plant will use more fuel than a tunnel kiln plant will. However, it is much more flexible and smaller diameter pipes can be fired as fast as 42 hours cold to cold.

Illustration 117 shows a tunnel kiln designed for firing pipe with a 4 meter wide setting width. A kiln car for a shuttle kiln would look exactly the same as the tunnel kiln car except it would be 6 meters wide instead of four.

Illustration #117 – Tunnel Kiln Firing Clay Pipe. Courtesy of Swindell Dressler.

About 20 years ago some roller hearths were built to fire plain end pipe. The advantage of this firing technique was that the weight of the kiln cars did not have to be heated and the disadvantage being that they had to make only plain end pipe. However, today with the development of the super lightweight fiber post car, tunnel kilns using these light weight cars can fire as fast as a roller hearth and are not limited to plain end pipe only.

19. DINNERWARE

There are four types of dinnerware, namely semi-vitreous, which is also called earthenware and stoneware, porcelain, hotel china and fine china. A discussion of each type of dinnerware is as follows:

19.1 Semi-Vitreous Dinnerware

Semi-vitreous dinnerware is the type of dinnerware used for every day use, is less expensive to make than either porcelain hotel china or fine china. Semi-vitreous dinnerware does not require kiln furniture in the bisque firing which substantially reduces the manufacturing cost. Also, semi-vitreous dinnerware is usually only fired twice, whereas porcelain, hotel china and fine china are fired three or more times.

The most common practice with semi-vitreous dinnerware is to have a bisque firing at approximately 1250°C in a tunnel kiln firing any where between 8 and 20 hour cycle cold to cold, based on the size pieces being fired. In other words, small plates can be fired in 8 hours, large bowls and pitchers may take up to 20 hours. As previously stated no kiln furniture is normally used other than one shelf, because the plates can be stacked one on top of the other for the bisque firing.

With a two-fire operation the fired bisque ware is glazed and then decorated by painting on the decoration. The glazed and decorated dinnerware is fired in the glost firing on special kiln furniture that keeps the plates from touching each other. This firing is normally done in the 1100 to 1200° range in fast fire, low profile tunnel kilns firing on a cycle between 6 and 15 hours cold to cold. A small percentage of semi-vitreous dinnerware is decorated using decals in which case a third decoration fire would be required. Decoration firing for decals usually is done in a belt or roller hearth type kiln on a very fast cycle, between 2 and 4 hours, firing to approximately 760°C.

19.2 Porcelain Dinnerware

The most popular type of dinnerware made in Asia and Europe is hard porcelain. Porcelain dinnerware is fired three times with the bisque firing being only to 800–1000°C. Since the body is only partially vitrified in the bisque firing, no kiln furniture is required and the product can be stacked. The bisque firing is only high enough so the product can be handled. Typical bisque cycles are 8–20 hours in tunnel kilns.

The second firing is the glost firing to approximately 1250°C and is also done in tunnel kilns. The product is fired in kiln furniture which totally supports the piece. Typical firing cycles are 8–15 hours long. The kiln atmosphere is oxidizing up to 1000°C, reducing from 1000 to 1280°C, and oxidizing during the cooling. The vitrification of the body is completed in the glost firing along with the development of the glaze.

Porcelain dinnerware is usually decorated over glaze with the decoration firing around 760–800°C. Belt kilns and roller hearths are used with the cycle times being in the 2 to 4 hour category.

19.3 Hotel China

Since hotel china is designed to take the abuse of restaurant use, the body is a high strength type of body that is fully vitrified. The plate cross section and weight is increased for strength purposes and the decorations are almost always underglazed. To make hotel china it takes three or more firings, the first firing being a bisque firing to around 1250°C, firing cycle time in the 8 to 20 hours range, based on the size of the pieces being fired. Unlike semi-vitreous, hotel china must be supported with kiln furniture in the bisque firing, to prevent slumping when complete vitrification takes place. This firing is done in tunnel kilns.

In the hotel china manufacturing process the second firing is the decoration firing which is commonly done with decals in a belt type kiln which does not require kiln furniture. Typical decorating temperature is 850°C and a typical heating time through the belt kiln would be two to four hours.

The last firing for hotel china is the glost firing normally done in tunnel kilns with the product being supported on cranks or special kiln furniture which prevents the glazed dinnerware from touching each other. Glost firings are typically in the 1080°C range with throughput times of 6 to 15 hours.

19.4 Fine China

Fine China or Bone China is the unique product first developed in England in the mid 1800's. The English bone china product uses bone ash in the body which promotes glass formation such that when the body is matured, the glass content can be almost 70%, which in turn makes the body translucent. A duplicate product can also be produced by adding ground up glass to the body to create the desired translucency. Fine china is fired three or more times with overglazed decorating being the norm. Overglaze decorating makes it possible to have more brilliant colors and a wide variety of colors. Overglazed decorating also provides the opportunity to add metalic gold and silver to the dinnerware.

The firing of fine china is normally done in a tunnel kiln, firing to approximately 1250°C, on firing times between 8 and 20 hours, depending on the size of the pieces being fired. Bisque firing of fine china requires kiln furniture to give total support to prevent slumping. Pieces must be made oversize to allow for shrinkage, which occurs during the bisque firing. Hollow ware often must be fired with supports for handles, etc.

The second firing in fine china is the glost firing which normally takes place around 1080°C in tunnel kilns with a throughput time of 6 to 15 hours. The glost firing operation, of course, requires kiln furniture to support the ware so it cannot touch each other during firing. The overglaze decorations used in fine china can be hand painted or decals. Decoration firing normally take place in belt type kilns or roller hearths at temperatures of 850°C with throughput times of 2 to 4 hours.

19.5 Lead Free Glazes

There is a strong movement in the world to eliminate lead bearing glazes and decorations on dinnerware. Lead free glazes and decorations are available, however, they are fired at higher temperatures. This is not a problem for glost firing in a tunnel kiln which is capable of going to these high temperatures easily. However, it is a problem for typical belt type decorating kilns where the temperature limitation on the kiln is the stainless steel belt. This is having the effect that many of the stainless steel belt type decorating kilns may have to be replaced with roller hearths capable of higher temperatures required for lead free decoration.

19.6 The Most Popular Type of Kilns for Firing Dinnerware are:

A. *Low Profile Tunnel Kilns for Bisque and Glost*
   Illustration 118 shows a typical low profile tunnel kiln bisque firing semivitreous dinnerware. The temperature uniformity from top to bottom in a low setting height tunnel kiln is better than it is in a tunnel kiln with a high setting height. Therefore the firing cycle can normally be speeded up. The advent of high velocity burners have made it possible to build tunnel kilns with very wide setting widths so that when the setting height is reduced the overall length of the kiln does not have to be increased. The net result of all this is that the low setting height, wide setting width tunnel kilns can fast fire dinnerware with excellent temperature uniformity compared to the older type designs with square cross sections. This type of kiln is the most popular today for both bisque and glost firing.

B. *Belt Kilns or Roller Hearths for Decoration*
   For decoration firing which is done at much lower temperatures, and very fast cycles, Belt type kilns are the most popular because of their

low cost to build and the fact that they do not have kiln cars which must be heated. New developments with lead free decorations and glazes have elevated the decoration temperature above that which is practical for belt kilns and now roller hearths are becoming more popular for decoration with the lead free glazes.

Illustration #118 – Low Profile Tunnel Kiln Bisque Firing Semi-Vitreous Ware. Courtesy of Swindell Dressler.

20. ELECTRIC INSULATORS

20.1 High Voltage Electrical Porcelain

The High Voltage Electrical Porcelain Industry developed primarily in this century as the need for transporting high voltage electricity over long distances developed. As a result, the high voltage electrical porcelain industry is a mature industry in the developed countries and very much a growing industry in the developing countries. High voltage electrical insulators are made in a great variety of sizes with small insulators being 100mm in diameter by 100mm long, up to very large insulators, more than 1 meter in diameter and 4 meters long. Electric insulators are always glazed and require a high mechanical strength as well as high electrical resistivity.

20.2 Firing Problems

High voltage electrical porcelain is once fired with the body and glaze maturing together. The typical firing temperature is 1280°C and the firing cycle time is dependent on the size of the insulators being fired. Insulators up to 1-1/2 meters high can usually be fired on a 48 hour firing cycle cold to cold, whereas the largest insulators which may be 3 or 4 meters high, may have to be fired in a cycle as long as five days. Suspension insulators up to 300 mm in diameter can be fired in 26 to 28 hours.

Electrical porcelain fired in the United States and England are fired in an oxidizing atmosphere, whereas electrical porcelain fired throughout most of the rest of the world are fired with a reduction firing cycle. Reduction firing has the advantage that sulfates are reduced to sulfites which dissociates at temperatures while the pores of the porcelain are still open. Also, ferric oxide is reduced to ferrous oxide. The advantage of this is with these gas sources removed, the pore structure is improved which in turn gives superior dielectric strength. This is particularly important for dense solid core insulators used extensively in Europe. However, insulators can be produced with acceptable properties using either the oxidation or reduction firing process. Therefore, the reason that the two processes exist simultaneously in the world is primarily due to tradition. The reduction firing process does create different colored glazes due to reduction of iron oxide. Therefore, it is unlikely that the Europeans with reduction fire would change that process or that the Americans with oxidation fire would change theirs.

Illustration 119 shows a typical insulator firing cycle for reduction firing large insulators up to 200 kg in weight. This is a 72 hour firing cycle where the atmosphere in the kiln is oxidizing up to about 1050°C in

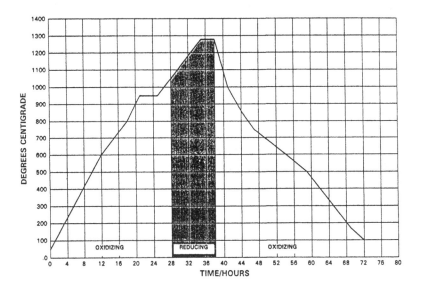

Illustration #119 – Typical Firing Cycle for Reduction Firing of Insulators.

order to facilitate burning out carbonaceous materials. At about 1050°C a strongly reducing atmosphere (4% CO) is introduced for approximately six hours up to a temperature of 1250°C. For the next two hours from 1250°C to 1275°C and one hour into the soak, the reduction environment is reduced to 2% CO. For the second hour of the soak, the reduction environment is reduced to 1% CO, after which for the last hour of the soak the atmosphere is neutral. The burners are turned off at the end of the soak period and cooling takes place under oxidizing conditions. For oxidation firing of the same size of insulators, the firing cycle profile is almost identical, with the only difference being that there is no reduction period, and the atmosphere stays oxiding throughout the entire cycle.

20.3 The Most Popular Type of Kilns for Firing High Voltage Electrical Porcelain are:

A. *Tunnel Kilns for Insulators up to 1.5 Meters High*

A tunnel kiln running on a 48 hour cold to cold firing cycle can fire all diameter insulators up to a height of 1.2 to 1.5 meters. Smaller insulators would be fired on shelving or on a refractory kiln furniture grid. Larger insulators up to 1.5 meters high would be fired vertically. This is an ideal application for a tunnel kiln since everything can be fired to the same temperature on the same time/temperature profile.

B. *Updraft Shuttle Kilns for Larger Insulators*

Large insulators, 1.5 meters to 4 meters tall, are set vertically in the kiln and are best fired in updraft shuttle kilns where the heating gases travel up through the load. If insulators are not uniformly heated around their perimeter they can bend.

Downdraft shuttle kilns tend to pull the heating gases from the outside of the load to the center which can differentially heat one side of the insulator versus the other. With updraft shuttle kilns, particularly where the burners are underfiring the load on an oscillating flame front basis. The heating gases rise uniformly up through the load heating all sides of the insulators at the same time.

21. FERRITES

21.1 Soft and Hard Ferrites

Soft ferrites are used for antenna rods, electric magnets, memory cores and transformer cores. Hard ferrites, on the other hand, are permanent magnets and they are used extensively in fractional horse power motors and very heavily in the automotive industry.

21.2 Firing Problems

Hard ferrites are fired in an air atmosphere to temperatures in the 1350 to 1400°C range. The cycle times vary depending on the largest piece being fired, but generally in the range of 16–40 hours cold to cold. Temperature uniformity is quite important with ferrites. Temperature spreads in the 1–2°C range are desired at the end of the soak period.

Soft ferrites also require a very tight temperature spread at the end of the soak period, and are fired to temperatures in the 1350 to 1400°C range. Firing cycle times for soft ferrites are also in the 16–40 hour range cold to cold, depending on the size of the pieces being fired. However, this is where the similarity ends between soft and hard ferrites. Soft ferrites must be fired in a air atmosphere during heat up and soak and then in a nitrogen atmosphere or more importantly, in the absence of oxygen for the cooling period. This requirement for an oxygen free cooling requires a very special type of pusher-plate kiln with water cooled jackets surrounding the load in order to control the cooling.

21.3 The Most Popular Types of Kilns for Ferrites are:

A. *Gas Fired Tunnel Kilns for Hard Ferrites*

Small cross section gas fired tunnel kilns are the most economical way to fire hard ferrites. The combination of the gas circulation within the tunnel kiln and the relatively small cross section produces very uniform heating conditions in the hot zone which is required. Hard ferrites are also fired in pusher plate kilns, however, pusher plate kilns generally cost more to build per cubic meter of output than tunnel kilns do. Since hard ferrites have no special atmospheric requirements, the more costly pusher plate design is not required. Illustration 120 shows a typical small cross section gas fired tunnel kiln for firing hard ferrites.

Illustration #120 – Tunnel Kiln Firing Hard Ferrites. Courtesy of Harrop Industries, Inc.

B. *Electric Pusher Plate Kilns with a Nitrogen Atmosphere in the Cooling Zone for Soft Ferrites*

Pusher plate kilns have the advantage of a gas tight outer shell completely surrounding the product on all four sides. The product is pushed through the kiln on slabs or refractory plates which run on refractory rails built into the kiln. Electric heating systems are normally used in a pusher plate which makes it easier to maintain the integrity of the atmosphere since there is no exhaust through the kiln. The temperature uniformity in an electric pusher plate kiln is also very good because of the small cross section which allows the radiation heat transfer to penetrate the entire load. Illustration 121 shows an electric pusher plate kiln firing soft ferrites.

Illustration #121 – Electric Pusher Plate Firing Soft Ferrites. Courtesy of Harrop Industries, Inc.

Recently batch type elevator kilns have been developed in Europe for firing soft ferrites. These kilns have an electric heating system and are designed to cool with nitrogen circulated through a heat exchanger to withdraw the heat. However, by far the largest majority of soft ferrites are still fired in electric pusher plate continuous kilns.

22. POTTERY

22.1 Types of Pottery

There are generally three types of pottery that are made, the first type being earthenware with a red body, fired to cone 6, (1200°C). The second type of pottery that is popular is stoneware which has a grey to a buff body color and is fired to cone 10 (1285°C). The third type of pottery is vitrious porcelain which is fired anywhere from cone 8 to 14 (1230–1350°C) and has a white body.

22.2 Firing Problems

Pottery is usually fired twice, with the first firing being a bisque firing to approximately 1000°C where the body is fired just hard enough to handle it. The second firing, which is a glost firing with decoration is done to the higher temperature which matures the body and develops the glaze and decoration. Most pottery is overglaze decorated where the decoration is painted on. Preparing the body for the glost fire, floor wax is often painted on the feet of the piece where no glaze is required. The floor wax burns off in the firing and provides a dry foot. Most pottery is made in relatively small shops which lends itself to relatively small batch fired kilns. Popular sizes of pottery kilns are 1/3 of a cubic meter to 1 cubic meter of usable setting space inside the kiln. These batch fired pottery kilns are usually the box type with a fixed hearth or a small shuttle kiln with a single car and a door built onto the end of the car. Firing cycle times are typically

24 hours cold to cold with the heating period being 10–12 hours. Recently reduction firing of pottery has become very popular because it produces warmer colors, which has caused a resurgence in gas fired pottery kilns, as compared to electric. As much as 50% of all pottery is now reduction fired.

22.3 The Most Popular Types of Pottery Kilns are:

A. *Gas Fired Box or Shuttle Kilns for Reduction Firing*

Gas fired pottery kilns make it easy to reduction fire since the reducing atmosphere can be controlled by simply adjusting the fuel air ratio in the kiln. The trend toward reduction firing of pottery has made gas fired pottery kilns, as shown in Illustration 122, very popular.

Illustration 122 – Gas Fired Pottery Kilns. Courtesy of Swindell Dressler.

B. *Electric Box or Shuttle Kilns for Oxidation Firing*

Most pottery that is fired in an oxidation atmosphere is done in an electric kiln of the box or shuttle type. A Typical electric fired kiln for firing pottery is shown in Illustration 123. Electric kilns are usually used for oxidation firing because they are less expensive than gas fired kilns and since no special atmosphere is required, they are the most economic.

Illustration #123 – Electric Box Type Pottery Kiln. Courtesy of Unique/Pereny.

23. REFRACTORIES

23.1 Types of Refractories that are Fired

There are four basic groups of refractories that are fired, group one being fireclay and high alumina bricks and shapes, group two are basic refractories, group three are silica refractories and group four are super refractories which is a catchall phrase that includes silicon carbide, mullite, 90 to 99% alumina, MgO, zircon and zirconia. Sometimes a large refractory plant might make refractories from more than one group. However, it is more common for a plant to be dedicated to fire clay and high alumina or to silica, or to basic brick, or to super refractories.

23.2 Firing Problems

A. *Fireclay and High Alumina*

Fireclay and high alumina plants are often located near the site of the fireclay mine. A typical fireclay may have an alumina content in the 33–38% range, and bauxite additions would be made so that various grades of refractories can be produced. For example, if the alumina content in the clay coming out of the ground was 35%, then a small amount of calcined alumina would be added to raise the alumina content to 38% in order to make the high duty fireclay brick. Larger percentages of calcined alumina would be used to upgrade to super duty, with approximately 45% alumina, or to produce the 50, 60, 70 and 80% alumina grades.

Generally speaking, it is the rule to fire refractories at or above their use temperature so that fireclay refractories are normally fired between 1300 and 1480, depending on their grade. High alumina refractories are fired in the range between 1450 and 1540. Firing cycle time for fireclay and high alumina depends mostly on the size of the pieces being fired. Standard brick shapes are normally fired in 48 hours cold to cold, up to several equivalents in size. Larger size shapes are fired in three to four day long cycles depending on size.

Refractory brick are normally fired without kiln furniture, with the green bricks stacked on top of each other. The setting height in the kiln is therefore determined by the maximum amount of weight that can be placed on the green bricks so that during the firing process the bottom brick is not deformed. With fireclay and high alumina the typical setting height is usually 10 to 12 courses high (1.4 meters) without kiln marking.

B. Basic Brick

This category includes magnesite, chrome, mixtures of chrome and magnesite and dolomite. Standard chrome and magnesite bricks are fired in the temperature ranges of 1450 to 1650°C and direct bonded chrome magnesite brick is fired in the temperature ranges of 1700–1800°C. Dolomite is normally fired in the ranges of 1450–1480°C.

Firing cycle times for basic brick are typically in the 48–50 hour range, cold to cold, in both tunnel kilns and batch kilns. As with other refractory materials the firing cycle time is more dependent on the physical size of the pieces being fired than it is on the material being fired. Basic brick is almost 50% more dense than fireclay and as a result, the setting heights in kilns can not be as high as they are with fireclay. To avoid kiln markings, setting heights in basic tunnel kilns are typically 600 mm and in batch kilns, 1 meter high. One of the problems encountered in basic brick tunnel kilns is that volatiles coming out of the product can cause severe alkali attack in the preheat zone. Therefore, it is common practice to have the preheat zone lined with a basic brick to resist this corrosive attack.

Illustration 124 shows a typical tunnel kiln firing cycle for firing direct bonded basic brick to 1800°C.

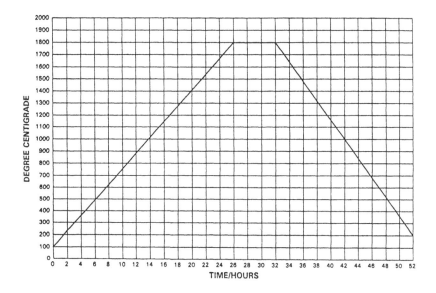

Illustration #124 – Typical Tunnel Kiln Cycle for Direct Bond Basic Brick.

C. *Silica Brick*

The most common use of pure silica refractory today is as coke oven blocks. Coke oven blocks come in a variety of sizes and the firing cycles vary in length of time depending on the size blocks being fired. However, the silica cycles are long to begin with, due to the volume change which occurs during the silica inversions in both heating and cooling. Illustration 125 shows three cycles for firing silica coke oven block in a shuttle kiln. Cycle #1 is 150 hours long and is used for firing blocks up to 15 kg in weight. Cycle #2 is 192 hours long and is used for firing blocks up to 30 kgs. Cycle #3 which is 237 hours in total length, is used for firing blocks larger than 30 kgs.

You will note that all three silica cycles have the same basic shape in that there is relatively slow heating below 600°C and slow cooling below 600°C during the cooling part of the cycle. Above 600°C the heating and cooling rates are much faster. The reason for this is that the silica inversion, alpha to beta quartz has a large volume change and if the individual pieces are not at uniform temperature at the time of the conversion the pieces will crack. After the conversion, the temperature uniformity requirement is not nearly as stringent and the heating rates and cooling rates can be accelerated. The relatively long soak periods are a function of the large sizes of the individual coke oven blocks.

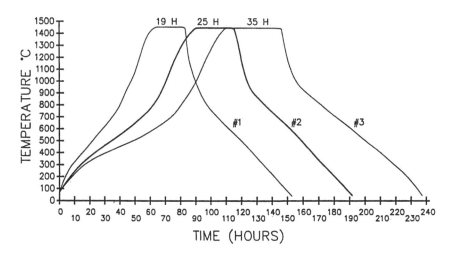

Illustration #125 – Firing Cycles for Silica Coke Oven Block Fired in a Shuttle Kiln.

D. Super Refractories

As previously stated, super refractories include silicon carbide, mullite, 90–99% alumina, MgO, zircon and zirconia. The great majority of super refractories are sold as either kiln furniture or for insulation in kilns and furnaces that operate at temperatures higher than where you can use standard fireclay and high alumina refractories.

The majority of super refractories are made as shapes so that the firing cycle time is more dependent on the size of the shape than it is on the particular material being fired. Firing cycle times for average size shapes and kiln furniture plates and posts is in the area of 48 hours cold to cold for any of these materials. However, large blocks used in the glass industry may have cycles as long as 10 to 15 days. Most super refractories are fired in an oxidizing atmosphere with the exception of nitride bonded silicon carbide, which is fired in retorts in a nitrogen atmosphere. Firing cycle temperatures for the various super refractory products are as follows:

Silicon Carbide	1400–1510°C.
Mullite	1600–1650°C.
90–99% Alumina	1500–1800°C.
MgO	1500–1800°C.
Zircon	1650–1680°C.
Zirconia	1700–1800°C.

23.3 The Most Popular Types of Kilns for Refractories are:

A. *Tunnel Kilns Plus Some Batch Kilns for Standard Size Fireclay and Basic Brick*

Standard size fireclay, high alumina and basic brick made in large quantities for the steel industry, can best be fired in tunnel kilns; however, most refractory plants also have to make shapes. Rather than slow the tunnel kiln cycle down to accommodate a small percentage of large shapes, it is usually advantageous to have bell kilns or shuttle kilns in the plant along with the tunnel kiln, to handle larger shapes, special orders and varying temperatures. In other words, the ideal situation is to keep the tunnel kiln in the refractory plant firing as much as possible on one firing schedule, to one temperature, which would accommodate the majority of the throughput of the plant, and let all fluctuations be handled in the batch firing kilns.

B. *Downdraft Shuttle Kilns for Silica Brick*

Silica brick and coke oven block are better fired in large downdraft shuttle kilns than in tunnel kilns for two reasons. Number one, firing cycles are so long with the silica coke oven blocks that the tunnel kiln, limited in setting height due to temperature uniformity, becomes extremely long. Silica holds its strength right up to its melting point and can be stacked quite high, such as two meters high. A two meter high setting height in a shuttle kiln is practical because the top burners can be zoned separately from the bottom burners to reduce top to bottom temperature spread, whereas in a tunnel kiln a two meter high setting height would automaticlly create an unsatisfactory temperature spread top to bottom. The second reason why shuttle kilns are better than tunnel kilns for silica coke oven block is the fact that there is a drastic change in the rate of heating before and after the quartz inversions. This big change in heating rate can be easier handled in a batch kiln than in a tunnel kiln.

Illustration #126 – Bell Kilns in a Super Refractory Plant. Courtesy of Swindell Dressler.

C. *Bell Kilns for Super Refractories*

Most Super refractory plants make a variety of products on a variety of firing cycles and in a wide variety of sizes. It is not uncommon for a single super refractories plant to have eight or ten different firing cycles. Also, super refractories are quite expensive and individual orders tend to be smaller in size than standard brick orders. Due to the fact that the variety of firing cycle requirements and production runs tend to be smaller, several bell kilns would be the ideal firing system for most super refractory plants. Illustration #126 shows a typical kiln installation in a super refractory plant, using multiple bell kilns.

24. SANITARYWARE

24.1 Sanitaryware First Fire and Refire

Sanitaryware which are porcelain bathroom fixtures are normally once fired in a 10–16 hour firing cycle in a continuous kiln to a temperature in the 1210–1220°C range. After the first firing approximately 80% of the pieces will be first class ware ready to be shipped to customers and 20% will not. Of the first fire rejects approximately 5% would be losses that are not repairable, usually from clay cracks, whereas 15% can be repaired. Of the 15% that can be repaired, the defects are clay cracks that can be repaired, or glaze defects such as dry spots, or too thick glaze, too thin glaze or dirt. After the repairs are made the pieces are refired, usually on a time temperature schedule of 12 to 24 hours. Of 100 pieces refired approximately 10% would be lost and another 5% would have to be repaired and refired again, with 85% shippable as first quality ware. The refiring is normally done in a shuttle kiln rather than a continuous kiln because the amount of product that must be refired varies quite a bit from week to week.

24.2 The Most Popular Types of Kilns for Sanitaryware are:

A. *Low Profile Tunnel Kilns for First Fire*

The most popular type of kiln for first firing sanitaryware today is the low profile tunnel kiln which is essentially one deck high with a high/ low set. Illustration 127 shows a low profile tunnel kiln for first firing sanitaryware. In this type of tunnel kiln the setting height is approximately 800 mm, the overall length of the kilns are in the 50 to 100 meter range, and setting widths are in the 2-1/2 to 3-1/2 meter range.

Roller hearths have also been used to first fire sanitaryware, however, they are severely limited in their capacity by their setting width limitation. Sanitaryware roller hearths usually have a setting width of 1-1/2 meters which means that they have to be quite a bit longer than a tunnel kiln to get the same capacity.

Illustration #127 – Tunnel Kiln for First Firing Sanitaryware. Courtesy of Swindell Dressler.

B. *Low Profile Shuttle Kilns for Refire*

The latest thinking in refire shuttle kilns favors a single deck kiln with a high/low setting exactly the same as in the tunnel kiln. Single deck shuttle kilns are extremely wide with a setting width in the 5 to 6.5

meter range. Illustration 128 shows a shuttle kiln for refiring sanitaryware with a 6.5 meter setting width. All fiber construction makes this type of shuttle kiln practical with either pulse firing or oscillating flame front under firing and an updraft exhaust. The uniformity in this type of kiln is excellent due to the low setting height and the ability to load and unload is comparable to a tunnel kiln with the same low profile.

Illustration #128 – 6.5 Meter Wide Shuttle Kiln for Refiring Sanitaryware. Courtesy of Swindell Dressler.

25. SPARK PLUGS

25.1 Spark Plug Bodies

Most spark plug bodies are made from 96% alumina and are fired in the 1600–1700°C range on 24 hour cold to cold firing cycle. Since spark plug cores are small in size, they could be fired faster than 24 hours, however, the firing cycle is really a function of firing the sagger to give good sagger life rather than a function of firing the core itself. After the high temperature firing of the alumina spark plug core, the spark plug is glazed and glost fired in the 800 to 1000°C range. This firing is often done in specialized equipment designed specifically for glost firing spark plug cores.

25.2 The Most Popular Type of Kiln for Firing Spark Plugs is a Small Cross Section Tunnel Kiln

Traditionally, spark plugs have been fired in small cross section tunnel kilns. A typical spark plug tunnel kiln would have a setting width of 600 mm, a setting height of 600 mm and an overall length of 30 to 35 meters. Since the firing temperature is in the 1600–1700°C range most spark plug tunnel kilns use preheated combustion air in some form to improve efficiency and elevate flame temperatures. Spark plug tunnel kilns are usually over and under fired continuous push with the saggers supported by refractory girders. The most popular designs use sidewall plenum chambers, where high velocity premix burners, inspirate and burn preheated combustion air from the cooling zone. Illustration 129 shows some typical spark plug tunnel kilns the preheated air type, just described.

Illustration #129 – Small Cross Section Tunnel Kilns Firing Spark Plugs. Courtesy of Swindell Dressler.

26. STRUCTURAL CLAY PRODUCTS

26.1 Face Brick, Roof Tile, Masonry Shapes

Structural clay product kilns are seldom sold alone as a kiln, but most commonly as part of a complete manufacturing plant. Therefore, the kiln must match the manufacturing process. Structural clay products are by and large building materials and fall into three major categories being face brick, roof tile and masonry shapes.

There are three processes for forming face brick, namely molded brick, where the brick is formed in molds with approximately 24% water. Secondly is soft mud extrusion where the brick is extruded through a dye with approximately 20% water and lastly, stiff mud extrusion where the extrusion is done with approximately 14 to 16% water. In the case of molded brick and soft mud brick, the brick is too soft to be stacked on kiln cars until after it is dried, therefore, stick dryers are used to dry the brick before stacking on kiln cars. The major advantage of stiff mud extrusion is that the bricks can be set on kiln cars prior to being dried and can be conveyed through the dryer and the kiln on the same kiln car.

There are two types of ceramic roof tile, namely European style and Japanese style. The shape and size of these two styles of roof tiles determine the types of kilns that are most efficient for firing them. For the European style, the roof tiles are fired on kiln furniture that holds the tile in the horizontal position. (H-Setter tile support system) With the tiles stacked 16 high. On the other hand, with the large Japanese roof tiles they are fired one high, standing on edge, which eliminates the need for kiln furniture and only requires pins to help stabilize the load.

The third class of structural clay products are masonry shapes, including wall and ceiling blocks. Illustration 130 shows a continuous push, over and under fired tunnel kiln, firing ceiling blocks.

Illustration #130 – Tunnel Kiln Firing Masonry Shapes. Courtesy of Swindell Dressler.

26.2 Firing Problems

One of the most difficult aspects of the structural clay products business is that the products must be manufactured very economically in order to be competitive. Economics dictate that the raw material be used almost as is, as it comes out of the ground with very little additives. Raw material for structural clay products are red iron bearing clays which vary considerably from place to place. Assuming a traditional setting height of 14 bricks high (1300 mm) firing cycle time can vary from 30 to 60 hours depending on the clay being used. Before the kiln can be designed the firing cycle must be determined. To determine the firing cycle curve, tests are usually made on the clay including DTA, TGA thermal expansion plus drying, firing and shrinkage tests. With this test data, a firing time/ temperature curve can be determined.

There is another major factor which effects the firing time and that is the setting pattern. For example, with a more traditional setting of 14 high in a bung which is set 11 over 4, the firing cycle might be 30 hours cold to cold. If the setting bung is reduced in size to 8 high in a package with only 5 over 2 setting, firing cycle time might be reduced to 20 hours. However in the two high concept where the setting package is actually two bricks, one set on top of the other, the firing cycle time could be reduced to as little as six to ten hours cold to cold.

A high percentage of structural clay products are flashed in the kiln which means that at the end of the soak period, during the early cooling the kiln atmosphere is changed to strongly reducing. This is accomplished by injecting raw gas into the tunnel which causes the Fe_2O_3 red iron oxide to change to Fe_3O_4, which is black in color. Flashing can create a whole variety of colors going from red to dark red, to garnet, to purple, to brown and black. The biggest problem with flashing is obtaining consistency of flash throughout the setting package. The smaller the setting package the less this is a problem.

26.3 The Most Popular Type of Kilns for Structural Clay are:

A. *The Most Popular Type of Kilns for Face Brick are Large, Wide Tunnel Kilns*

The most popular type of kilns for face brick are large wide tunnel kilns, however, there are three types within this category. These kilns can be top fired index push, side fired index push, or under and over fired with continuous push. All these systems work. However, traditionally the top fired kilns have been the most popular in Europe, and underfired continuously pushed kilns have been the most popular in the United States. These wide face brick tunnel kilns are quite commonly made with setting widths from 6 meters to 9 meters wide.

Since the tunnel kiln is only part of the brick manufacturing plant the overall maufacturing process which produces the product for the minimum cost will be the most popular kiln system for the future. The debate over what is the most economical manufacturing system for face brick now centers over the size of the setting package that is fired in the kiln.

As previously stated, the firing cycle time can be drastically reduced by making the firing package smaller. For example, if the firing package is 14 high with a setting of 11 over 4, the cycle time might be 30 hours, also the drying time might be 30 hours for a total of 60 hours. If that firing package is reduced to 8 high with a setting of 5 over 2, the firing and drying times may be reduced to 20 hours each (40 hours total). Illustration 131 shows a tunnel kiln with an 8 high firing package.

Illustration #131 – Facebrick Tunnel Kiln (Firing Package 8 High). Courtesy of Swindell Dressler.

The new two high concept can use a package of one brick standing on end as a soldier course or two bricks set on edge, one on top of the other; or three bricks flat set on top of each other, for a total setting height in the kiln of 230 mm. The firing time can be reduced down to 10 or less hours and the drying time can also be reduced a similar amount. If the drying time is reduced from 30 to 10 hours and the firing time is reduced from 30 to 10 hours then the total drying and firing time is reduced from 60 hours to 20 hours, which facilitates just in time manufacturing, faster deliveries and lower inventories.

Two-high plants offer other benefits besides shorter drying and firing times. There are significant savings in electric power especially in drying. Also there is significant savings in hacking and de-hacking machinery from the capital cost standpoint, as well as the operating cost standpoint, and the maintenance standpoint. The reason for this is the equipment for hacking and de-hacking with only two high is very simple and relatively inexpensive. Lastly, two high plants have shown that flashing is much more consistent because of the smaller firing package.

B. *The Most Popular Type of Kilns for Roof Tiles are:*

(1) *High Pressure Tunnel Kilns for European Style*
High pressure tunnel kilns for roof tile have been designed and perfected in Europe, some of which have water seals between the kiln cars and the kiln walls,to hold in the high pressure. This design is effective in creating temperature uniformity and low fuel consumption with roof tiles set in horizontal setters. The advantage of the high pressure is that the kiln can be built with very tight clearances to force the heating gases through the product load at relatively high velocities and therefore transfer heat in a more efficient manner.

(2) *One High Tunnel Kilns for Japanese Style*
The Japanese style roof tile has a quite different shape from the European style and they are large and can be set on edge, one high, on a tunnel kiln car that contains pins to give the roof tiles stability. This system eliminates most of the kiln furniture which does not have to be fired and simplifies automatic loading and unloading of the kiln cars.

27. WALL AND FLOOR TILE

27.1 Types of Wall and Floor Tiles

During the last 25 years the firing of floor and wall tile has changed dramatically. Twenty five years ago glazed ceramic wall tile was fired in kiln furniture called cranks in conventional tunnel kilns on firing cycles of 20 to 24 hours in duration. Today, these same type of wall tiles are fired in cycles as short as 30 minutes in roller hearth kilns which are actually a part of a continuous manufacturing line which starts with the press and ends with the packaging machine. Today it is not untypical for the elapsed time between the pressing of the tile and the time when the finished glazed fired tile is placed into a box for shipment is no more than 90 minutes. Almost every size tile today, glazed or unglazed, are best fired in roller hearths. One exception to this could be unglazed mosaic tiles, 50 mm by 50 mm, which can be fired in saggers on their edge, very efficiently, in low profile tunnel kilns. It is common to fire glazed tile as small as 100 by 100 mm directly on the rolls of a roller hearth kiln with no kiln furniture being required. Larger size tiles such as 150 mm by 150 mm, 200 mm by 200 mm, and 300 mm by 300 mm have been fired in roller hearth kilns without the use of kiln furniture for years.

27.2 Firing Problems

The Italians who make more ceramic tile than anyone else in the world are primarily responsible for this revolution in the manufacturing process of tile. The Italian companies have perfected the roller hearth firing techniques by not only developing better equipment, but by engineering the ceramic bodies so that they could be fired as fast as is capable with roller hearth type of kilns. Illustration 132 shows a 30 minute firing cycle for wall tile of the 150 mm by 150 mm size. These amazingly short firing cycles are only possible with specially engineered bodies and glazes that can mature in this short period of time.

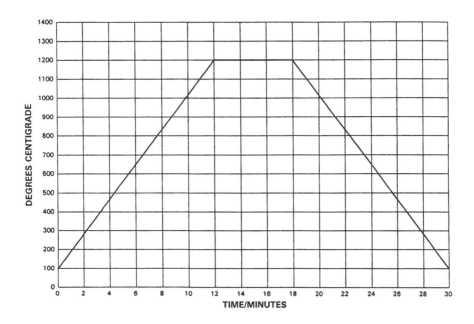

Illustration #132 – 30 Minute Firing Cycle — Wall Tile.

27.3 The Most Popular Type of Kiln for Floor and Wall Tiles are Roller Hearths

Today, very few roller hearths kilns are sold by themselves for firing tile. Normally the roller hearth kiln is sold as part of a complete manufacturing line that starts with a press and ends with the packaging machine. Roller hearth kilns for firing tile are built in lengths of 35 to 110 meters, with usable setting widths on the rolls between one and two meters. Two channel roller hearths are sometimes built with two sets of rolls, one on top of the other, in order to increase capacity within a given floor space. However, the trend today favors a one set of rolls or a single channel roller hearth design, as handling machinery is less complicated.

Illustration 133 shows a typical roller hearth for firing wall tile.

Illustration #133 – Roller Hearth For Wall Tile. Courtesy of S.I.T.I. S.p.A.

APPENDIX

Melting Points of Some Compounds and Minerals

	Deg. C.	Deg. F
Alumina	2050	3722
Andalusite	1816	3301
Andalusite (commercial)	1655	3011
Arsenous oxide	200	392
Barium carbonate	1360	2480
Barium chloride	960	1760
Barium oxide	(O_2) 450	(O_2) 842
Barium sulfate	1580	2876
Bauxite	2035	3695
Bauxite (commercial)	1800-2020	3272-3668
Borax	Red Heat	
Calcite	2570	4658
Calcite (commercial)	2095-2485	3803-4505
Calcium carbonate (dissociates)	825	1517
Calcium fluoride	1300	2372
Calcium oxide	2570	4658
Calcium sulfate (gypsum) dissociates	900	1652
Chromium oxide	2330	4226
Cobaltic oxide	(O_2) 905	(O_2) 1661
Cobalt nitrate	56	133
Copper oxide (Cu_2O)	1210	2210
Copper oxide (CuO)	1064	1947
Corundum	2035	3695
Corundum (commercial)	1850-2030	3362-3686
Cyanite	1816	3301
Cyanite (commercial)	1680	3056
Diaspore	2035	3695
Diaspore (commercial)	1920	3488
Diatomaceous earth	1715	3119
Diatomaceous earth (commercial)	1650	3002
Dolomite	2570-2800	4658-5072
Dolomite (commercial)	1925-2485	3497-4505
Ferric oxide	1548	2818
Ferrous oxide	1419	2586
Fireclay (high grade)	1660-1720	3020-3128
Fireclay (low grade)	1600-1650	2912-3002
Flint	1715	3119
Fluorspar	1300	2372
Forsterite	1910	3470
Ganister	1715	3119
Gibbsite	2035	3695
Gibbsite (commercial)	1760-2030	3200-3686
Halloysite	1775	3227
Kaolin	1740-1785	3164-3245
Kaolinite	1785	3245
Kyanite	1820	3308
Lead oxide (Litharge)	880	1616
Lead oxide (Minimum), dissociates	500-530	932-986

	Deg. C.	Deg. F.
Lime	2570	4658
Limestone	2570	4658
Magnesite (dissociates)	2800	5072
Magnesite (commercial)	2000-2800	3632-5072
Magnesium carbonate (dissociates)	350	662
Magnesium oxide (approx.)	2825	5117
Magnetite	1538	2800
Manganese dioxide	(O_2) 1058	(O_2) 570
Mullite	1810	3290
Mullite (commercial)	1790	3254
Nickel oxide	(O_2) 400	(O_2) 752
Orthoclase feldspar (dissociates)	1170	2138
Potassium carbonate	880	1616
Potassium chromate	975	1787
Potassium dichromate	398	748
Potassium nitrate	337	639
Potassium oxide	Red Heat	
Quartz	1715	3119
Rutile (dissociates)	1900	3452
Rutile (commercial)	1630	2966
Silica	1715	3119
Silicon carbide (decomp.)	2200	3992
Sillimanite	1816	3301
Sillimanite (commercial)	1810	3290
Sodium carbonate	853	1567
Sodium chloride	792	1458
Sodium nitrate	313	595
Sodium oxide	Red Heat	
Sodium sulfate	880	1616
Spinel	2135	3875
Spinel (commercial)	1915	3479
Tin oxide	1130	2066
Titanium oxide (dissociates)	1900	3452
Whiting (dissociates)	825	1517
Zircon (dissociates)	2550	4622
Zircon (commercial)	1900-2300	3452-4172
Zirconia	2700	4892

Following are the melting points of some glass-forming silicates:

	Deg. C.	Deg. F.
Na_2SiO_3	1089	1992
K_2SiO_3	976	1789
$PbSiO_3$	770	1418
$BaSi_2O_5$	1426	2599
Beta $CaSiO_3$	1540	2804

Temperature scale conversions

to C	F or C	to F		to C	F or C	to F
−17.78	0	32		315.6	600	1112
−12.22	10	50		326.7	620	1148
−6.67	20	68		337.8	640	1184
−1.11	30	86		348.9	660	1220
0.00	32	90		360.0	680	1256
4.44	40	104		371.1	700	1292
10.00	50	122		398.9	750	1382
15.56	60	140		426.7	800	1472
21.11	70	158		454.4	850	1562
26.67	80	176		482.2	900	1652
32.22	90	194		510.0	950	1742
37.78	100	212		537.9	1000	1832
43.33	110	230		565.6	1050	1922
48.89	120	248		593.3	1100	2012
54.44	130	266		621.1	1150	2102
60.00	140	284		648.9	1200	2192
65.56	150	302		676.7	1250	2282
71.11	160	320		704.4	1300	2372
76.67	170	338		732.2	1350	2462
82.22	100	356		760.0	1400	2552
87.78	190	374		787.8	1450	2642
93.33	200	392		815.6	1500	2732
98.89	210	410		843.3	1550	2922
100.0	212	414		871.1	1600	2912
104.4	220	428		898.9	1650	3002
110.0	230	446		926.7	1700	3092
115.6	240	464		954.4	1750	3182
121.1	250	482		982.2	1800	3272
126.7	260	500		1010	1850	3362
132.2	270	518		1038	1900	3452
137.8	280	536		1066	1950	3542
143.3	290	554		1093	2000	3632
148.9	300	572		1121	2050	3722
154.4	310	590		1149	2100	3812
160.0	320	608		1177	2150	3902
165.6	330	626		1204	2200	3992
171.1	340	644		1232	2250	4082
176.7	350	662		1260	2300	4172
182.2	360	680		1288	2350	4262
187.8	370	698		1316	2400	4352
193.3	380	716		1343	2450	4442
198.9	390	734		1371	2500	4532
204.4	400	752		1427	2600	4712
215.6	420	788		1482	2700	4892
226.7	440	824		1538	2800	5072
237.8	460	860		1593	2900	5252
248.9	480	896		1649	3000	5432
260.0	500	932		1704	3100	5612
271.1	520	968		1760	3200	5792
282.2	540	1004		1816	3300	5972
293.3	560	1040		1871	3400	6152
304.4	580	1076		1927	3500	6332
				1982	3600	6512

Unit equivalents

Metric To American	American To Metric
Metric To Metric	American To American

AREA:

1 mm² = 0.001 55 in.² = 0.000 010 76 ft²
1 cm² = 0.155 in.² = 0.001 076 ft²
1 m² = 1550 in.² = 10.76 ft²

1 in.² = 645.2 mm² = 6.452 cm²
  = 0.000 645 2 m²
1 ft² = 92 903 mm² = 929.03 cm²
  = 0.0929 m²
1 acre = 43 560 ft²
1 circular mil = 0.7854 square mil
  = 5.067 × 10⁻¹⁰ m² = 7.854 × 10⁻⁴ in.²

DENSITY and SPECIFIC GRAVITY:

1 g/cm³ = 0.036 13 lb/in.³ = 62.43 lb/ft³
  = 1000 kg/m³ = 1 kg/*l*
  = 62.43 lb/ft³ = 8.345 lb/USgal
1 μg/m³ = 136 grains/ft³
  (for particulate pollution)
1 kg/m³ = 0.062 43 lb/ft³

1 lb/in.³ = 27.68 g/cm³ = 27 680 kg/m³
1 lb/ft³ = 0.0160 g/cm³ = 16.02 kg/m³
  = 0.0160 kg/*l*
Specific gravity relative to water
  (SGW) of 1.00 = 62.43 lb/ft³ at 4 C or 39.2 F†
Specific gravity relative to dry air
  (SGA) of 1.00 = 0.0765 lb/ft³‡
  = 1.225 kg/m³
1 lb/USgal = 7.481 lb/ft³ = 0.1198 kg/*l*
1 g/ft³ = 35.3 × 10⁶ μg/m³
1 lb/1000 ft³ = 16 × 10⁶ μg/m³

ENERGY, HEAT, and WORK:

1 cal = 0.003 968 Btu
1 kcal = 3.968 Btu = 1000 cal = 4186 J
  = 0.004 186 MJ
1 J = 0.000 948 Btu = 0.239 cal = 1 W·s
  = 1 N·m = 10⁷ erg = 10⁷ dyne·cm
1 W·h = 660.6 cal

3413 Btu = 1 kW·h
1 Btu = 0.2929 W·h
  = 252.0 cal = 0.252 kcal
  = 778 ft·lb
  = 1055 J = 0.001 055 MJ
1 ft·lb = 0.1383 kg·m = 1.356 J
1 hp·hr = 1.98 × 10⁶ ft·lb
1 therm = 1.00 × 10⁵ Btu
1 bhp (boiler horsepower) = 33 475 Btu/hr
  = 8439 kcal/h = 9.81 kW

HEAT CONTENT and SPECIFIC HEAT:

1 cal/g = 1.80 Btu/lb = 4187 J/kg
1 cal/cm³ = 112.4 Btu/ft³
1 kcal/m³ = 0.1124 Btu/ft³ = 4187 J/m³
1 cal/g·°C = 1 Btu/lb·°F = 4187 J/kg·°K

1 Btu/lb = 0.5556 cal/g = 2326 J/kg
1 Btu/ft³ = 0.008 90 cal/cm³
  = 8.899 kcal/m³ = 0.0373 MJ/m³
1 Btu/USgal = 0.666 kcal/*l*
1 Btu/lb·°F = 1 cal/g·°C = 4187 J/kg·°K

Unit equivalents *(concluded)*

Metric to American	American To Metric
Metric To Metric	American To American

VOLUME *(cont'd)*

1 Br gal = 277.4 in.^{3}
         = 0.004 546 m^3 = 4.546 l
         = 1.201 USgal
1 bbl, oil = 9702 in.^{3} = 5.615 ft^3
           = 0.1590 m^3 = 159.0 l
           = 42.00 USgal
           = 34.97 Br gal

VOLUME FLOW RATE:

1 cm^3 (cc)/s = 1 × 10^{-6} m^3/s
1 l/s = 1 × 10^{-3} m^3/s
1 m^3/h = 4.403 US gpm (gal/min)
          = 0.5887 ft^3/min

1 gpm (gal/min) = 60.0 gph (gal/hr)
                = 0.016 67 gps (gal/sec)
                = 0.002 23 cfs (ft^3/sec)
                = 0.1337 cfm (ft^3/min)
                = 0.8326 BR gpm
                = 0.227 m^3/h
                = 1.429 bbl/hr
                = 34.29 bbl/day
1 gph (gal/hr) = 0.0631 l/s
               = 0.000 037 1 cfs (ft^3/sec)
1 cfm (ft^3/min) = 6.18 Br gpm
                   = 0.000 471 m^3/s
1 cfs (ft^3/sec) = 448.8 gpm
                   = 22 250 Br gph

WEIGHT, FORCE, MASS:

1 g = 0.035 27 oz avdp mass
1 kg mass = 1000 g mass
          = 35.27 oz avdp mass
          = 2.205 lb avdp mass
1 kg force = 1000 g force = 9.807 N
           = 2.205 lb avdp force
1 metric ton = 1000 kg = 2205 lb

1 oz avdp mass = 28.35 g = 0.028 35 kg
1 lb avdp mass = 453.6 g = 0.4536 kg
               = 4.536 × 10^5 μg
1 lb avdp force = 0.4536 kg force
                = 4.448 N
1 lb = 7000 grains
1 short ton = 2000 lb = 907.2 kg
1 long ton = 2240 lb = 1015.9 kg

Unit equivalents *(continued)*

Metric To American
Metric To Metric

American To Metric
American To American

VISCOSITY, absolute, μ:

0.1 Pa·s = 1 dyne·s/cm² = 360 kg/h·m
   = 1 poise = 100 centipoise
   = 242.1 lb mass/hr·ft
   = 0.002 089 lb force·sec/ft²
1 kg/h·m = 0.672 lb/hr·ft = 0.002 78 g/s·cm
   = 0.000 005 81 lb force·sec/ft²

1 lb mass/hr·ft = 0.000 008 634
   lb force·sec/ft²
   = 0.413 centipoise
   = 0.000 413 Pa·s
1 lb force·sec/ft² = 115 800 lb mass
   = 47 880 centipoi
   = 47.88 Pa·s
1 reyn = 1 lb force·sec/in.²
   = 6.890 × 10⁴ centipoise

μ of water† = 1.124 centipoise
   = 2.72 lb mass/hr·ft
   = 2.349 × 10⁻⁵ lb·sec/ft²

μ of air† = 0.0180 centipoise
   = 0.0436 lb/hr·ft
   = 3.763 × 10⁷ lb·sec/ft

VISCOSITY, kinematic, ν:

1 cm²/s = 0.0001 m²/s
   = 1 stoke = 100 centistokes
   = 0.001 076 ft²/sec
   = 3.874 ft²/hr
1 m²/s = 3600 m²/h
   = 38 736 ft²/hr = 10.76 ft²/sec

1 ft²/sec = 3600 ft²/hr = 92 900 ce
   = 0.0929 m²/s
1 ft²/h = 0.000 278 ft²/sec = 25.8 ce
   = 0.000 025 8 m²/s

ν of water† = 1.130 centistokes
   = 32 SSU
   = 1.216 × 10⁻⁵ ft²/sec

ν of air† = 14.69 centistokes
   = 1.581 × 10⁻⁴ ft²/sec

VOLUME:

1 cm³ (cc) = 0.000 001 00 m³
   = 0.0610 in.³ = 0.0338 US fluid oz
1 l (dm³) = 0.0010 m³ = 1000 cm³
   = 61.02 in.³ = 0.035 31 ft³
   = 0.2642 USgal

1 in.³ = 16.39 cm³ = 0.000 163 9 m³
   = 0.016 39 l
1 ft³ = 1728 in.³ = 7.481 USgal
   = 6.229 Br gal
   = 28 320 cm³ = 0.028 32 m³ =
   = 62.427 lb of 39.4 F (4 C) wate
   = 62.344 lb of 60 F (15.6 C) wat

1 m³ = 1000 l = 1 000 000 cm³
   = 61 020 in.³ = 35.31 ft³
   = 220.0 Br gal
   = 6.290 bbl
   = 264.2 USgal
   = 1.308 yd³

1 USgal = 3785 cm³ = 0.003 785 m³
   = 3.785 l = 231.0 in.³
   = 0.8327 Br gal = 0.1337 ft³
   = sp gr × 8.335 lb
   = 8.335 lb of water
   = ¹/₄₂ barrel (oil)

Unit equivalents *(continued)*

Metric To American
Metric To Metric

American To Metric
American To American

PRESSURE (cont'd)

(For rough calculations,
1 bar = 1 atm = 1 kg/cm²
= 10 m H₂O = 100 kPa)

1 atm† = 101.3 kPa = 101 325 N/m²
= 10 330 mm H₂O = 407.3 in. H₂O
= 760.0 mm Hg = 29.92 in. Hg
= 235.1 oz/in.² = 14.70 lb/in.²
= 1.033 kg/cm²
= 1.013 bar

TEMPERATURE:

$C = \frac{5}{9}(F - 32)$
$F = \frac{9}{5}(C + 32)$
$K = C + 273.15$
$R = F + 459.67$

THERMAL CONDUCTIVITY:

1 W/m·°K = 0.5778 Btu·ft/ft²·hr·°F
= 6.934 Btu·in./ft²·hr·°F
1 cal·cm/cm²·s·°C = 241.9 Btu·ft/ft²·hr·°F
= 2903 Btu·in./ft²·hr·°F
= 418.7 W/m·°K

1 Btu·ft/ft²·hr·°F = 1.730 W/m·°K
= 1.488 kcal/m·h·°K
1 Btu·in./ft²·hr·°F = 0.1442 W/m·°K
1 Btu·ft/ft²·hr·°F = 0.004 139 cal·cm/cm²·s·°C
1 Btu·in./ft²·hr·°F = 0.000 344 5 cal·cm/cm²·s·°C

THERMAL DIFFUSIVITY:

1 m²/s = 38 760 ft²/hr
1 m²/h = 10.77 ft²/hr

1 ft²/hr = 0.000 025 8 m²/s = 0.0929 m²/h

VELOCITY:

1 cm/s = 0.3937 in./sec = 0.032 81 ft/sec
= 10.00 mm/s = 1.969 ft/min
1 m/s = 39.37 in./sec = 3.281 ft/sec
= 196.9 ft/min = 2.237 mph
= 3.600 km/h = 1.944 knot

1 in./sec = 25.4 mm/s = 0.0254 m/s
= 0.0568 mph
1 ft/sec = 304.8 mm/s = 0.3048 m/s
= 0.6818 mph
1 ft/min = 5.08 mm/s = 0.005 08 m/s
= 0.0183 km/h
1 mph = 0.4470 m/s = 1.609 km/h
= 1.467 ft/sec
1 knot = 0.5144 m/s
1 rpm = 0.1047 radians/sec

Unit equivalents (continued)

Metric To American	American To Metric
Metric To Metric	American To American

HEAT FLOW, POWER:

1 N·m/s = 1 W = 1 J/s
         = 0.001 341 hp = 0.7376 ft·lb/sec
1 kcal/h = 1.162 J/s = 1.162 W
         = 3.966 Btu/hr
1 kW = 1000 J/s = 3413 Btu/hr
       = 1.341 hp

1 hp = 33 000 ft lb/min = 550 ft lb/sec
      = 745.7 W = 745.7 J/s
      = 641.4 kcal/h
1 Btu/hr = 0.2522 kcal/h
          = 0.000 393 1 hp
          = 0.2931 W = 0.2931 J/s

HEAT FLUX and HEAT TRANSFER COEFFICIENT:

1 cal/cm²·s = 3.687 Btu/ft²·sec
            = 41.87 kW/m²
1 cal/cm²·h = 1.082 W/ft² = 11.65 W/m²
1 kW/m² = 317.2 Btu/ft²·hr
1 kW/m²·°C = 176.2 Btu/ft²·hr·°F

1 Btu/ft²·sec = 0.2713 cal/cm²·s
1 Btu/ft²·hr = 11.35 kW/m²
            = 2.713 kcal/m²·h
1 kW/ft² = 924.2 cal/cm²·h
1 Btu/ft²·hr·°F = 4.89 kcal/m²·h·°C

LENGTH:

1 mm = 0.10 cm = 0.039 37 in.
        = 0.003 281 ft
1 m   = 100 cm = 1000 mm = 39.37 in.
        = 3.281 ft
1 km = 0.6214 mile

1 in. = 25.4 mm = 2.54 cm = 0.0254 m
1 ft = 304.8 mm = 30.48 cm = 0.3048 m
1 mile = 5280 ft
1 micron = 1 μ = 10^{-6} m
1 Angstrom unit = 1 Å = 10^{-10} m

PRESSURE:

1 N·m² = 0.001 kPa = 1.00 Pa
1 mm H_2O = 0.0098 kPa
1 mm Hg = 0.1333 kPa = 13.60 mm H_2O
           = 1 torr = 0.0193 3 lb/in.²
1 kg/cm² = 98.07 kPa = 10 000 kg/m²
            = 10 000 mm H_2O = 394.1 in. H_2O
            = 735.6 mm Hg = 28.96 in. Hg
            = 227.6 oz/in.² = 14.22 lb/in²
            = 0.9807 bar
1 bar = 100.0 kPa = 1.020 kg/cm²
        = 10 200 mm H_2O = 401.9 in. H_2O
        = 750.1 mm Hg = 29.53 in. Hg
        = 232.1 oz/in.² = 14.50 lb/in.²
        = 100 000 N/m²
1 g/cm² = 0.014 22 lb/in.²
          = 0.2276 oz/in.²
          = 0.3937 in. H_2O

1 in. H_2O = 0.2488 kPa = 25.40 mm H_2O
           = 1.866 mm Hg
           = 0.002 54 kg/cm² = 2.54 g/cm²
1 in. Hg   = 3.386 kPa = 25.40 mm Hg
           = 345.3 mm H_2O = 13.61 in. H_2O
           = 7.858 oz/in.² = 0.491 lb/in.²
           = 25.4 torr
1 lb/in.²   = 6.895 kPa = 6895 N/m²
           = 703.1 mm H_2O = 27.71 in. H_2O
           = 51.72 mm Hg = 2.036 in. Hg
           = 16.00 oz/in.²
           = 0.0703 kg/cm² = 70.31 g/cm²
           = 0.068 97 bar
1 oz/in.² = 0.4309 kPa
           = 43.94 mm H_2O = 1.732 in. H_2O
           = 3.232 mm Hg
           = 0.004 39 kg/cm² = 4.394 g/cm²

Comparative data (by weight) for some typical fuels

	Heating value Btu/lb (and Btu/gal)		Heating value kcal/kg (and kcal/l)		Gross Btu per scf air[10]	Wt air req'd per unit wt fuel (and scf/gal)	Weight of combustion products per wt of fuel (and ft³/gal)				Ultimate vol % CO_2 in dry flue gas
	Gross	Net	Gross	Net			CO_2	H_2O	N_2	Total	
Blast furnace gas	1 179	1 079	665	599	135.3	0.57	0.58	0.01	1.08	1.67	25.5
Coke oven gas	18 595	16 634	10 331	9 242	104.4	13.63	1.51	1.81	8.61	11.93	10.8
Producer gas[1]	2 614	2 459	1 452	1 366	129.2	1.55	0.61	0.15	1.72	2.48	18.4
Natural gas[2]	21 830	19 695	12 129	10 943	106.1	15.73	2.55	2.03	12.17	16.75	11.7
Propane, natural	21 573 (91 500)	19 886 (84 345)	11 986 (6094)	11 049 (5617)	107.5	15.35 (850.8)	3.01 (108.11)	1.62 (144.39)	12.01 (682.06)	16.64 (934.57)	13.8
Butane, refinery	20 810 (102 600)	19 183 (94 578)	11 562 (6833)	10 658 (6299)	106.1	15.00 (949.0)	3.04 (124.27)	1.53 (146.92)	11.82 (747.18)	16.39 (1018.4)	14.3
Methanol	9 700 (64 150)	8 400 (55 550)	5 389 (4272)	4 667 (3700)	106.4	6.47 (559.5)	1.38 (78.4)	1.13 (156.8)	4.97 (445.3)	7.48 (681)	15.0
Gasoline. motor	20 190 (123 351)	18 790 (114 807)	11 218 (8216)	10 440 (7646)	104.6	14.80 (1183)	3.14 (165.1)	1.30 (166.8)	11.36 (940.3)	15.80 (1272)	15.0
#1 Distillate oil	19 423 (131 890)	18 211 (123 650)	10 791 (8784)	10 118 (8235)	102.1	14.55 (1292)	3.17 (185.7)	1.20 (171.0)	11.10 (1020)	15.48 (1377)	15.4
#2 Distillate oil	18 993 (137 080)	17 855 (128 869)	10 553 (9130)	9 920 (8583)	101.2	14.35 (1354)	3.20 (199.1)	1.12 (170.6)	10.95 (1070)	15.27 (1440)	15.7
#4 Fuel oil	18 844 (143 010)	17 790 (135 013)	10 470 (9524)	9 884 (8992)	103.0	13.99 (1388)	3.16 (206.7)	1.04 (166.1)	10.68 (1097)	14.92 (1472)	15.8
#5 Residual oil	18 909 (149 960)	17 929 (142 190)	10 506 (9987)	9 961 (9470)	104.2	13.88 (1439)	3.24 (221.0)	0.97 (161.4)	10.59 (1137)	14.81 (1520)	16.3
#6 Residual oil	18 126 (153 120)	17 277 (145 947)	10 071 (10 198)	9 599 (9720)	103.2	13.44 (1484)	3.25 (236.4)	0.84 (149.0)	10.25 (1172)	14.36 (1558)	16.7
Wood, non-resinous	6 300		3 500		98.4	4.90	1.39	0.65	3.47	5.51	20.3
Coal, bituminous	14 030		7 795		99.3	10.81	2.94	0.49	8.26	11.71	18.5
Coal, anthracite	12 680		7 045		97.8	9.92	2.96	0.22	7.58	10.78	19.9
Coke	12 690		7 051		96.2	10.09	3.12	0.07	7.73	10.94	20.4

Illustration #51 – Comparative Data by Weight for typical fuels. Courtesy of Combustion Handbook by North American Mfg. Co.

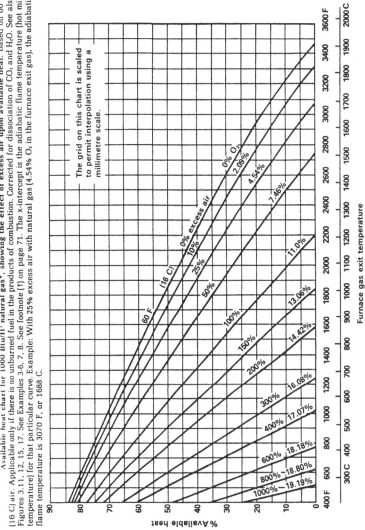

Available heat chart for 1000 Btu/ft³ natural gas*, showing the effect of excess air upon available heat. Based on 60 F (16 C) air. Applicable only if there is no unburned fuel in the products of combustion. Corrected for dissociation of CO_2 and H_2O. See also Figures 3.11, 12, 15, 17. See Examples 3-6, 7, 8. See footnote (†) on page 71. The x-intercept is the adiabatic flame temperature (hot mix temperature) for that particular curve. Example: With 25% excess air with natural gas (4.54% O_2 in the furnace exit gas), the adiabatic flame temperature is 3070 F, or 1668 C.

The grid on this chart is scaled to permit interpolation using a millimetre scale.

% Available heat

Furnace gas exit temperature

Illustration #53 – Effect of Excess Air on Available Heat. Courtesy of Combustion Handbook by North American Mfg. Co.

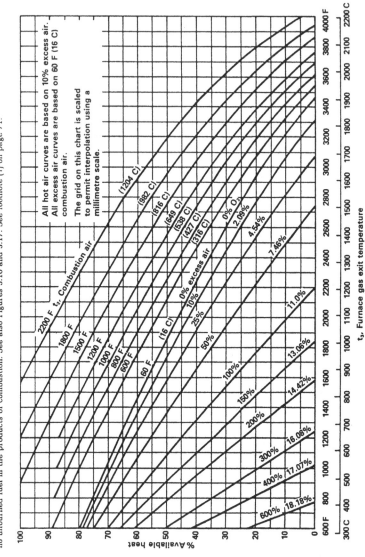

Available heat for 1000 Btu/ft³ natural gas with preheated combustion air at 10% excess air. Applicable only if there is no unburned fuel in the products of combustion. See also Figures 3.10 and 3.17. See footnote (†) on page 71.

All hot air curves are based on 10% excess air.
All excess air curves are based on 60 F (16 C) combustion air.

The grid on this chart is scaled to permit interpolation using a millimetre scale.

Illustration #54 – Effect of Preheated Combustion Air on Available Heat and Flame Temperature. Courtesy of Combustion Handbook by North American Mfg. Co.

Fuel savings resulting from use of preheated air with natural gas and 10% excess air. These figures are for evaluating a proposed change to preheated air—not for determining system capacity. Read captions for Tables 3.16b and c.

% Fuel saved with natural gas, 10% XSAir	t_2, Combustion air temperature, F													
t_1, Furnace gas exit temperature, F	600	700	800	900	1000	1100	1200	1300	1400	1500	1600	1800	2000	2200
1000	13.4	15.5	17.6	19.6	—	—	—	—	—	—	—	—	—	—
1100	13.8	16.0	18.2	20.2	22.2	—	—	—	—	—	—	—	—	—
1200	14.3	16.6	18.7	20.9	22.9	24.8	—	—	—	—	—	—	—	—
1300	14.8	17.1	19.4	21.5	23.6	25.6	27.5	—	—	—	—	—	—	—
1400	15.3	17.8	20.1	22.3	24.4	26.4	28.4	30.2	—	—	—	—	—	—
1500	16.0	18.5	20.8	23.1	25.3	27.3	29.3	31.2	33.0	—	—	—	—	—
1600	16.6	19.2	21.6	24.0	26.2	28.3	30.3	32.2	34.1	35.8	—	—	—	—
1700	17.4	20.0	22.5	24.9	27.2	29.4	31.4	33.4	35.3	37.0	38.7	—	—	—
1800	18.2	20.9	23.5	26.0	28.3	30.6	32.7	34.6	36.5	38.3	40.1	—	—	—
1900	19.1	21.9	24.6	27.1	29.6	31.8	34.0	36.0	37.9	39.7	41.5	44.7	—	—
2000	20.1	23.0	25.8	28.4	30.9	33.2	35.4	37.5	39.4	41.3	43.0	46.3	—	—
2100	21.2	24.3	27.2	29.9	32.4	34.8	37.0	39.1	41.1	43.0	44.7	48.0	51.0	—
2200	22.5	25.7	28.7	31.5	34.1	36.5	38.8	40.9	42.9	44.8	46.6	49.9	52.8	—
2300	24.0	27.3	30.4	33.3	36.0	38.5	40.8	42.9	45.0	46.9	48.7	52.0	54.9	57.5
2400	25.7	29.2	32.4	35.3	38.1	40.6	43.0	45.2	47.2	49.2	51.0	54.2	57.1	59.7
2500	27.7	31.3	34.7	37.7	40.5	43.1	45.5	47.7	49.8	51.7	53.5	56.8	59.6	62.2
2600	30.1	33.9	37.3	40.5	43.4	46.0	48.4	50.6	52.7	54.6	56.4	59.6	62.4	64.9
2700	33.0	37.0	40.6	43.8	46.7	49.4	51.8	54.0	56.1	58.0	59.7	62.8	65.5	67.9
2800	36.7	40.8	44.5	47.8	50.8	53.4	55.8	58.0	60.0	61.9	63.5	66.5	69.1	71.3
2900	41.4	45.7	49.5	52.8	55.7	58.4	60.7	62.8	64.7	66.4	68.0	70.8	73.2	75.2
3000	47.9	52.3	56.0	59.3	62.1	64.6	66.7	68.7	70.4	72.0	73.5	75.9	78.0	79.8
3100	57.3	61.5	65.0	68.0	70.5	72.7	74.6	76.2	77.7	79.0	80.2	82.2	83.8	85.2
3200	72.2	75.6	78.3	80.4	82.2	83.7	85.0	86.1	87.1	87.9	88.7	89.9	90.9	91.8

Illustration #55 – Fuel Savings with Preheated Air. Courtesy of Combustion Handbook by North American Mfg. Co.

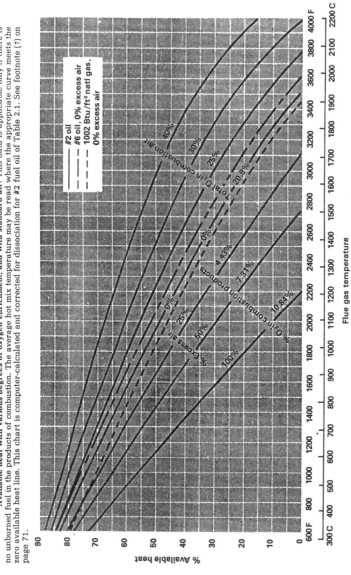

Available heat with various degrees of oxygen enrichment, and with standard air. This data is applicable only if there is no unburned fuel in the products of combustion. The average hot mix temperature may be read where the appropriate curve meets the zero available heat line. This chart is computer-calculated and corrected for dissociation for #2 fuel oil of Table 2.1. See footnote (†) on page 71.

Illustration #57 Effect of Oxygen Enrichment on Available Heat and Flame Temperature. Courtesy of Combustion Handbook by North American Mfg. Co.

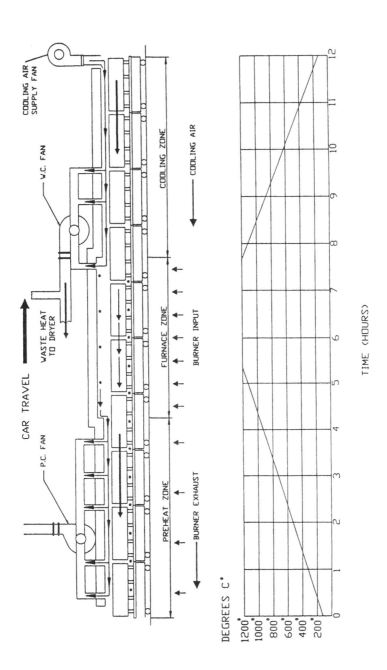

Illustration #81 – Heat Transfer in a Continuous Kiln.

INDEX